Advanced Numerical Methods with Matlab® 2

Mathematical and Mechanical Engineering Set

coordinated by
Abdelkhalak El Hami

Volume 7

Advanced Numerical Methods with Matlab® 2

Resolution of Nonlinear, Differential and Partial Differential Equations

Bouchaib Radi
Abdelkhalak El Hami

iSTE

WILEY

First published 2018 in Great Britain and the United States by ISTE Ltd and John Wiley & Sons, Inc.

ISTE Ltd
27-37 St George's Road
London SW19 4EU
UK

www.iste.co.uk

John Wiley & Sons, Inc.
111 River Street
Hoboken, NJ 07030
USA

www.wiley.com

Library of Congress Control Number: 2018934991

British Library Cataloguing-in-Publication Data
A CIP record for this book is available from the British Library
ISBN 978-1-78630-293-9

MIX
Paper from responsible sources
FSC® C013604

Contents

Preface

Most physical problems can be expressed in the form of mathematical equations (e.g. differential equations, integral equations). Historically, mathematicians had to find analytic solutions to the equations encountered in engineering and related fields (e.g. mechanics, physics, biology). These equations are sometimes highly complex, requiring significant work to be simplified. However, in the mid-20th Century, the introduction of the first computers gave rise to new methods for solving equations: numerical methods. This new approach allows us to solve the equations that we encounter (when constructing models) as accurately as possible, thereby enabling us to approximate the solutions of the problems that we are studying. These approximate solutions are typically calculated by computers using suitable algorithms.

Practical experience has shown that, compared to standard numerical approaches, a carefully planned and optimized methodology can improve the speed of computation by a factor of 100 or even higher. This can transform a completely unreasonable calculation into a perfectly routine computation, hence our great interest in numerical methods! Clearly, it is important for researchers and engineers to understand the methods that they are using and, in particular, the limitations and advantages associated with each approach. The computations needed by most scientific fields require techniques to represent functions as well as algorithms to calculate derivatives and integrals, solve differential equations, locate zeros, find the eigenvectors and eigenvalues of a matrix, and much more.

The objective of this book is to present and study the fundamental numerical methods that allow scientific computations to be executed. This involves implementing a suitable methodology for the scientific problem at hand, whether derived from physics (e.g. meteorology, pollution) or engineering (e.g. structural mechanics, fluid mechanics, signal processing).

This book is divided into two parts, with two appendices. The first part contains two chapters dedicated to solving nonlinear equations and differential equations. The second part consists of four chapters on the various numerical methods that are used to solve partial differential equations: finite differences, finite elements, finite volumes and meshless methods.

Each chapter starts with a brief overview of relevant theoretical concepts and definitions, with a range of illustrative numerical examples and graphics. At the end of each chapter, we introduce the reader to the various *Matlab* commands for implementing the methods that have been discussed. As is often the case, practical applications play an essential role in understanding and mastering these methods. There is little hope of being able to assimilate them without the opportunity to apply them to a range of concrete examples. Accordingly, we will present various examples and explore them with *Matlab*. These examples can be used as a starting point for practical exploration.

Matlab is currently widely used in teaching, industry and research. It has become a standard tool in various fields thanks to its integrated toolboxes (e.g. optimization, statistics, control, image processing). Graphical interfaces have been improved considerably in recent versions. One of our appendices is dedicated to introducing readers to *Matlab*.

Bouchaib RADI
Abdelkhalak EL HAMI
March 2018

PART 1

Solving Equations

1

Solving Nonlinear Equations

1.1. Introduction

Let $F\colon \mathbb{R} \to \mathbb{R}$ be a function defined on a subset $D \subset \mathbb{R}$. Suppose that we wish to find the roots of the equation $F(x) = 0$, if they exist. Most theorems that guarantee the existence of roots for equations of the form $F(x) = 0$ do not specify a way to construct these roots, and we are usually not capable of solving the problem analytically, except in a few special cases. Therefore, we often need to resort to numerical methods to find approximate solutions of the equation $F(x) = 0$ [RAD 09, BAK 76].

There are many possible numerical methods to choose from, and each has its own advantages and disadvantages (we will only present the most common methods in this chapter). Before trying to solve a problem numerically, we need to know that the solution exists and is unique [QUA 04, RAD 10]. Therefore, the process of numerically solving the equation $F(x) = 0$ can be divided into two parts:

1) First, we must show the existence of real-valued solutions and separate each solution that we wish to approximate in an interval $[a, b] \subset D$.

2) Second, we must choose a numerical method to approximate the isolated roots.

1.2. Separating the roots

DEFINITION.– *We say that the root $\bar{x}$ of the equation $F(x) = 0$ is separable if there exists an interval $[a, b] \subset D$, such that $\bar{x}$ is the only root contained in $[a, b]$.*

If so, the root $\bar{x} \in [a, b]$ is said to be separated or isolated. In practice, we can separate the roots of the equation $F(x) = 0$ either graphically (by looking for the points at which the graph of F intersects with the $(o\vec{x})$ axis) or analytically, by applying the intermediate value theorem (see the appendix on standard results from analysis, Appendix 2).

EXAMPLE.–

1) The equation $x^3 - 6x + 2 = 0$ has three real roots that are separated by the intervals $[-3, -2]$, $[0, 1]$ and $[2, 3]$.

2) The equation $e^x \sin x - 1 = 0 \Leftrightarrow \sin x = e^{-x}$ has two roots that are separated by the intervals $[0, 1]$ and $[3, 4]$.

3) The equation $x \ln x - 1 = 0$ has a single root in $[1, 2]$.

4) The equation $2x^4 + 3x^3 - 4x - 5 = 0$ has two real roots separated by the intervals $[-2, -1]$ and $[1, 2]$.

5) The equation $(1+x)e^{1-x} - 3/2 = 0$ has two real roots separated by the intervals $[-1, 0]$ and $[1, 2]$.

1.3. Approximating a separated root

In this section, we will assume that the root $\bar{x}$ has already been isolated by the interval $[a, b]$.

We will present some of the standard methods that can be used to find an approximate value for $\bar{x}$, and we will study their properties and rates of convergence. Concretely, when we speak of approximating $\bar{x}$ up to ϵ, we mean finding an approximate value $\tilde{\bar{x}}$ that is known to satisfy $|\bar{x} - \tilde{\bar{x}}| \leq \epsilon$ (where ϵ is a small real number; the smaller this ϵ, the more precise the algorithm).

1.3.1. *Bisection method (or dichotomy method)*

Suppose that $\bar{x}$ is a simple separable root in $[a, b]$ $(F(a)F(b) < 0)$. The bisection method constructs a sequence of intervals $I_n = [a_n, b_n]$ from the initial interval $[a, b]$, such that each I_n contains $\bar{x}$ and has a width equal to half of the width of I_{n-1}.

The underlying idea of the method is to construct the following three sequences (a_n), (b_n), and (x_n):

Define $a_0 = a$; $b_0 = b$, $x_0 = \frac{1}{2}(a_0 + b_0)$, and for $n \geq 1$:

– If $F(a_{n-1})F(x_{n-1}) < 0$, define $a_n = a_{n-1}$, $b_n = x_{n-1}$, $x_n = \frac{1}{2}(a_n + b_n)$.

– If $F(a_{n-1})F(x_{n-1}) > 0$, define $a_n = x_{n-1}$, $b_n = b_{n-1}$, $x_n = \frac{1}{2}(a_n + b_n)$.

– If $F(a_{n-1})F(x_{n-1}) = 0$, then $\bar{x} = x_{n-1}$: this means that we have found the exact solution, hence the process is terminated.

In this way, at each iteration, we select the half-interval that contains the root $\bar{x}$. After n iterations, we have found an interval containing $\bar{x}$ with width $\dfrac{b-a}{2^n}$.

THEOREM.– *The sequence* (x_n) *constructed by the bisection algorithm converges to the root* $\bar{x}$.

An upper bound for the error committed by approximating $\bar{x}$ *by* x_n *is given by:* $|x_n - \bar{x}| \leq \dfrac{b-a}{2^{n+1}}$.

REMARK.–

1) Similarly, the sequences (a_n) and (b_n) converge to $\bar{x}$, and $|\bar{x}-a_n| \leq (b-a)/2^n$; $|\bar{x} - b_n| \leq (b-a)/2^n$.

2) To guarantee a precision of ϵ when approximating $\bar{x}$ by x_n, we can simply pick n, such that: $n \geq \ln\left[\dfrac{(b-a)}{\epsilon}\right] / \ln 2 - 1$.

This method has the following properties:

– straightforward algorithm (based on the intermediate value theorem);

– can be applied to non-analytic functions;

– cannot be applied to double roots;

– cannot be applied to multiple equations;

– very slow (to achieve high precision, we need a high number of iterations).

Owing to these limitations, we need other methods in some situations.

EXAMPLE.– Suppose that we wish to find a zero of the following function using the bisection method:

$$f(x) = x\sin(x) - 1.$$

This function satisfies $f(0) = -1$, $f(2) = 0.818595$. It is continuous on $[0, 2]$, and $f(0)f(2) < 0$. By the intermediate value theorem, the equation $f(x) = 0$ has at least one root $\alpha \in [0, 2]$. We can therefore use this interval to initialize the bisection method. Applying this method to f on the interval $[0, 2]$, we find:

i	x_i	i	x_i
1	1.000000	7	1.109375
2	1.500000	8	1.117188
3	1.250000	9	1.113281
4	1.125000	10	1.115234
5	1.062500	11	1.114258
6	1.093750	12	1.114258

An approximate value for the root of f is therefore given by $x_{12} = 1.114258$.

1.3.2. *Fixed-point method*

DEFINITION.– *The solutions of the equation $f(x) = x$ are said to be the fixed points of the function f.*

THEOREM.– *If f is continuous on $[a, b]$ and $f([a, b]) \subset [a, b]$, then f has at least one fixed point between a and b (in other words, $\exists c \in]a, b[\ / \ f(c) = c$).*

Suppose that $\bar{x} \in [a, b]$ is an isolated root of the equation $F(x) = 0$. The fixed-point method defines and studies a function f such that $\bar{x}$ is a fixed point of f (i.e. such that $F(x) = 0 \Leftrightarrow f(x) = x$).

EXAMPLE.– For example, for the function $F(x) = (1 + x)e^{1-x} - \frac{3}{2}$, we can define $f(x) = \frac{3}{2}e^{x-1} - 1$.

Now, instead of searching for $\bar{x}$ as a root of $F(x)$, we can find it as a fixed point of $f(x)$ by taking advantage of the following property:

If the recursive sequence (x_n) defined by x_0 arbitrary; $x_n = f(x_{n-1})$, $n \geq 1$, converges, then its limit is a fixed point of the function f (see Figure 1.1).

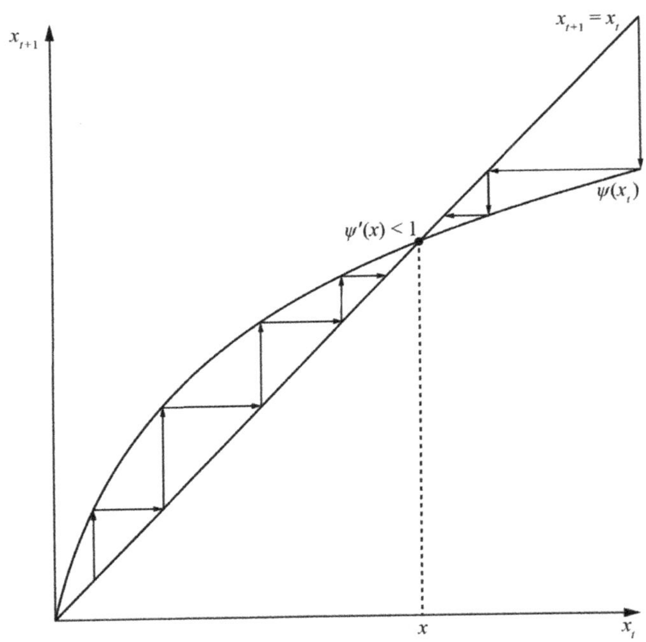

Figure 1.1. *Fixed-point method*

The idea of the method is to choose a function f for the original function F, such that:

– $F(x) = 0 \to f(x) = x$.

– The sequence (x_n) defined by $x_0 \in [a, b]$; $x_n = f(x_{n-1})$ for $n \geq 1$ converges to $\overline{x}$.

The problem of approximating $\bar{x}$ up to ϵ is therefore reduced to that of computing x_n, such that $|x_n - \bar{x}| \leq \epsilon$.

REMARK.– For any given function $F(x) = 0$, there are an infinite number of choices of f such that $F(x) = 0$ is equivalent to $f(x) = x$ (e.g. $f(x) = x + \lambda F(x)$; $\lambda \in \mathbb{R}^*$). The challenge of the fixed-point method lies in choosing f in such a way that the sequence (x_n) converges, and does so as quickly as possible. There are two natural criteria for selecting the best function: convergence and rate of convergence. The choice of the initial point x_0 also influences both of these criteria.

EXAMPLE.– Suppose that we wish to compute the value of $\sqrt{2}$. This is equivalent to finding the positive root $\alpha = \sqrt{2}$ of the function

$$f(x) = x^2 - 2,$$

that is, solving a nonlinear equation.

It is easy to check that $\alpha = \sqrt{2}$ is a fixed point of the function $\phi(x) = -\frac{1}{4}x^2 + x + \frac{1}{2}$ simply by noting that $\phi(\sqrt{2}) = \sqrt{2}$. Moreover, $\forall x \in [1, 2]$, $|\phi'(x)| \leq \frac{1}{2} = C$. We can therefore define the sequence $x_0 \in [1, 2]$ and $x_{n+1} = \phi(x_n)$, and apply the mean value theorem to deduce that $\exists \eta \in [1, 2]$, such that

$$|x_{n+1} - \alpha| = |\phi(x_n) - \phi(\alpha)| = |\phi'(\eta)||x_n - \alpha|.$$

Therefore,

$$|x_n - \alpha| \leq C^n |x_0 - \alpha|, \ \forall n \geq 0.$$

This shows that the sequence x_n converges to the root α. We need to perform 34 iterations of the fixed-point method to compute an approximate value for $\sqrt{2}$ that is accurate to ten digits after the decimal point.

1.3.3. *First convergence criterion*

THEOREM.– *Let $\overline{x} \in [a, b]$ be an isolated root of $F(x)$, and suppose that $f(x)$ is a continuous and differentiable function on $[a, b]$ satisfying $f(\overline{x}) = \overline{x}$.*

If there exists a real number k, $0 < k < 1$, such that:

i) $f([a, b]) \subset [a, b]$;

ii) $\forall x \in [a, b], |f'(x)| \leq k$;

then the recursive sequence (x_n) *defined by* $x_0 \in [a, b]; x_n = f(x_{n-1})$ *converges to* $\overline{x}$.

An upper bound for the error committed by approximating $\overline{x}$ *with* x_n *is given by:*

$$|\overline{x} - x_n| \leq k^n |\overline{x} - x_0|.$$

1.3.4. *Iterative stopping criteria*

The idea of this method is to find $\overline{x}$ by computing the limit of a convergent numerical sequence. In practice, we need to find an approximation of $\overline{x}$ up to a certain accuracy ϵ in a finite number of iterations (as few as possible). As $\bar{x}$ is unknown, we cannot directly apply the stopping criterion $|\overline{x} - x_n| \leq \epsilon$.

1.3.4.1. *Choosing the number of iterations*

To guarantee an accuracy of ϵ when approximating $\bar{x}$ by x_n, we can simply choose the smallest integer n, such that $k^n|b - a| \leq \epsilon$. Let $n \geq \ln\left(\frac{\epsilon}{b-a}\right) / \ln k$.

The smaller the number k, the faster the convergence.

The condition $k^n|b - a| \leq \epsilon$ is an extremely strong condition that is sufficient but not necessary! Before we can apply this condition, we need to know the value of k. If this is not possible, we will need an alternative stopping condition.

1.3.4.2. *Testing the absolute value*

In some cases, we can use a stopping condition based on the absolute value of the difference between two consecutive approximations of $\overline{x}$, which can be expressed in the form of the inequality $|x_n - x_{n-1}| \leq \epsilon$.

Of course, this condition is not sufficient to guarantee that x_n approximates the exact solution $\bar{x}$ up to ϵ, since $|x_n - x_{n-1}| \leq \epsilon$ does not imply that $|x_n - \overline{x}| \leq \epsilon$ in general. Whether or not the latter inequality holds depends on how the sequence (x_n) converges to $\bar{x}$.

We can write $|x_n - \bar{x}|$ as a function of $|x_n - x_{n-1}|$ as follows:

CASE 1.– $-1 < f'(x) < 0$ on $[a, b]$.

In this case, the sequences (x_{2n}) and (x_{2n+1}) are adjacent and their limit $\bar{x}$ is always located between any two consecutive terms of the sequence (x_n). Therefore, in this specific case, $|x_n - \bar{x}| \leq |x_n - x_{n-1}|$, and the stopping condition $|x_n - x_{n-1}| \leq \epsilon$ is sufficient to guarantee that x_n and x_{n-1} approximate $\bar{x}$ up to ϵ, one from above and the other from below. Therefore, we can approximate $\bar{x}$ up to ϵ without needing to find k.

CASE 2.– $0 < f'(x) < 1$ (or if the sign of f' is not known on $[a, b]$).

In this case, the sequence (x_n) converges monotonically, and we can show that

$$|x_n - \bar{x}| \leq \frac{1}{1-k}|x_{n+1} - x_n|.$$

Therefore, to guarantee a precision of ϵ when approximating $\bar{x}$ by x_n, we can simply apply the stopping condition $|x_{n+1} - x_n| \leq \epsilon(1 - k)$.

1.3.5. *Second convergence criterion (local criterion)*

THEOREM.– *Let $\bar{x} \in [a, b]$ be an isolated root of $F(x)$, and let $f(x)$ be a C^1 function, such that $f(\bar{x}) = \bar{x}$.*

1) If $|f'(\bar{x})| < 1$, then there exists a neighborhood V of $\bar{x}$ such that the recursive sequence (x_n) associated with f converges to $\bar{x}$, $\forall x_0 \in V$.

Furthermore, (x_n) converges monotonically if $0 \leq f'(\bar{x}) < 1$, and converges while oscillating around $\bar{x}$ if $-1 < f'(\bar{x}) < 0$.

The set V is called the domain of attraction of $\bar{x}$.

2) If $|f'(\bar{x})| > 1$, then, $\forall x_0$, the sequence $x_n = f(x_{n-1})$ does not converge to $\bar{x}$.

3) If $|f'(\bar{x})| = 1$, then we cannot draw any conclusions about the convergence behavior of the sequence (x_n), which is unstable (either convergent or divergent) and very sensitive to rounding errors.

REMARK.–

1) The convergence criterion $|f'(\bar{x})| < 1$ only holds in the neighborhood of $\bar{x}$ (we might not have convergence for every x_0 in $[a, b]$). We therefore need to choose the right x_0 (in the domain of attraction) to apply this criterion.

2) The smaller the value of $|f'(\bar{x})|$, the faster the convergence ($|f'(\bar{x})|$ is known as the rate of convergence).

3) In order to use this criterion, we need to compute $f'(\bar{x})$. This is not always possible (since $\bar{x}$ is an unknown in the problem).

The properties of this method can be summarized as follows:

– mostly only useful in theoretical contexts;

– constructive method;

– need a strategy to choose f and x_0.

1.3.6. *Newton's method (or the method of tangents)*

Suppose that we have an isolated root $\bar{x}$ of the equation $F(x) = 0$ in the interval $[a, b]$, and suppose that $F'(x)$ does not vanish on $[a, b]$.

Newton's method chooses the following specific function f to study the roots of F (see Figure 1.2):

$$f(x) = x - \frac{F(x)}{F'(x)}. \qquad [1.1]$$

Clearly, $\bar{x}$ is a fixed point of f. What can we say about the recursive sequence (x_n) defined by $x_0 \in [a, b]$, $x_{n+1} = f(x_n)$?

THEOREM.– *Let $\bar{x}$ be an isolated root of the equation $F(x) = 0$ in $[a, b]$. Suppose that F has continuity class C^2 and that F' does not vanish on $[a, b]$.*

Then there exists a neighborhood V of $\bar{x}$ on which the sequence (x_n) constructed by Newton's method converges to $\bar{x}$, $\forall x_0 \in V$.

THEOREM.– *Let $\bar{x}$ be an isolated root of the equation $F(x) = 0$ in $[a, b]$. Suppose that F' does not vanish on $[a, b]$, and that F'' is continuous on $[a, b]$. Let $M = \max|F''(x)|/2$ on $[a, b]$. If $x_0 \in [a, b]$ and $|x_0 - \bar{x}| < \frac{1}{M}$, then the sequence (x_n) constructed by Newton's method converges to the solution $\bar{x}$.*

An upper bound for the error committed by approximating $\bar{x}$ with x_n is given by:

$$|x_n - \bar{x}| \leq \frac{1}{M}(M|x_0 - \bar{x}|)^{2^n}.$$

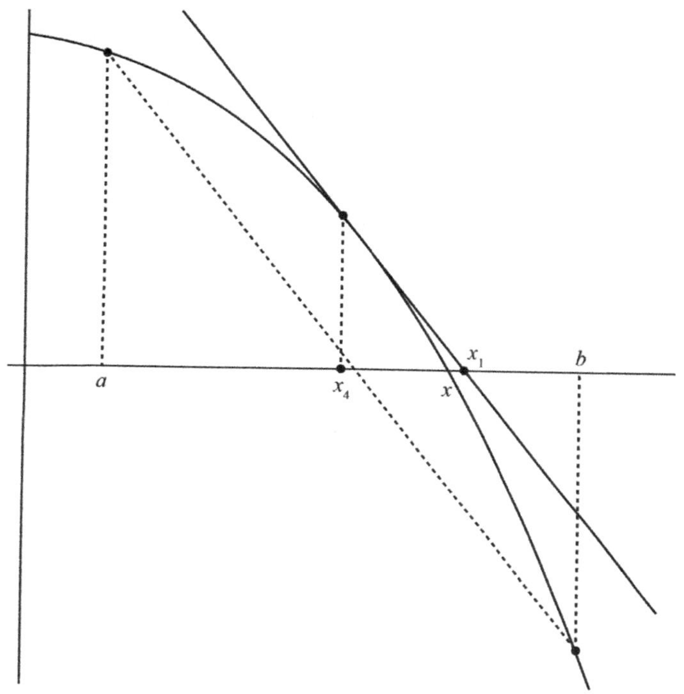

Figure 1.2. *Newton's method*

REMARK.–

1) Note that $f(\bar{x}) = 0$, so convergence is guaranteed on some neighborhood V of $\bar{x}$.

2) The convergence condition of Newton's process is expressed in terms of the choice of initial value $|x_0 - \bar{x}| < \frac{1}{M}$. This condition is satisfied $\forall x_0 \in [a, b]$ whenever $|b - a| < \frac{1}{M}$.

Therefore, the challenge of Newton's method is to choose the right x_0.

3) The following expression gives an alternative upper bound for the error:

$$|x_n - \bar{x}| \leq \frac{1}{L}(L|x_0 - \bar{x}|)^{2^n},$$

where $L = \max \left| \dfrac{F''(x)}{2F'(x)} \right|$.

4) Theorem 1.3 gives a sufficient condition on the choice of x_0 to guarantee convergence, but this condition is not necessary. The assumptions of the theorem

are often difficult to satisfy (since $\bar{x}$ is unknown). Therefore, in practice, we work backwards, as follows:

We define the sequence $x_{n+1} = f(x_n)$, then pick an arbitrary $x_0 \in [a, b]$; if this sequence converges, the limit is the desired solution; otherwise, we choose another x_0 in $[a, b]$.

5) Newton's method achieves quadratic convergence, which is faster than the previous method.

1.3.7. *Secant method*

In cases where we cannot compute the value of $F'(x)$ explicitly, we can instead use the approximation:

$$\frac{F(x_n) - F(x_{n-1})}{x_n - x_{n-1}}. \qquad [1.2]$$

This transforms the recurrence formula into:

$$x_{n+1} = x_n - F(x_n)\frac{x_n - x_{n-1}}{F(x_n) - F(x_{n-1})}, \quad x_0, x_1 \in [a, b]. \qquad [1.3]$$

Conceptually, we are replacing the tangent to the curve $F(x)$ at the point $(x_n, F(x_n))$ with the straight line that passes through this point and the point $(x_{n-1}, F(x_{n-1}))$.

EXAMPLE.– In 1225, Leonardo di Pisa studied the equation

$$f(x) = x^3 + 2x^2 + 10x - 20 = 0 \qquad [1.4]$$

and found the solution $x \simeq 1.368808107$. Nobody knows how he found his result, but it is an astonishing achievement given the state of knowledge at the time.

There are several ways in which to rewrite equation [1.4] in the form $x = F(x)$. If we define

$$x = \frac{20}{x^2 + 2x + 10} = F(x),$$

then

$$\lim_{x \to +\infty} f(x) = +\infty \quad \lim_{x \to -\infty} f(x) = -\infty.$$

By the intermediate value theorem, there exists at least one root on $]-\infty, +\infty[$. Computing the derivative of the function f gives

$$\begin{aligned} f'(x) &= 3x^2 + 4x + 10 \\ &= (\sqrt{3}x + \frac{2}{\sqrt{3}})^2 + \frac{26}{3}. \end{aligned}$$

Therefore, $f'(x) > 0$ for every $x \in \mathbb{R}$, so the equation $f(x) = 0$ has a unique root in $\mathbb{R}$. More precisely, $f(1) = -7$ and $f(2) = 16$ so the unique root of f on $\mathbb{R}$ is contained in the interval $[1, 2]$.

Hence, to find this root, we can simply restrict attention to the interval $[1, 2]$. The solution satisfies

$$x = \frac{20}{x^2 + 2x + 10} = F(x) = \frac{20}{(x+1)^2 + 3^2}.$$

It can easily be shown that the image of $[1, 2]$ under F is $[1, 2]$. The derivative of F is

$$F'(x) = \frac{-40(x+1)}{((x+1)^2 + 9)^2},$$

so, for all $x \geq 0$,

$$|F'(x)| = \frac{40(x+1)}{((x+1)^2 + 9)^2} \leq K < 1.$$

We define

$$g(x) = \frac{40(x+1)}{((x+1)^2 + 9)^2}.$$

After performing a few calculations, we find that

$$g'(x) = \frac{120((x+1)^2 + 9)(3 - (x+1)^2)}{((x+1)^2 + 9)^4}.$$

The following table shows the variations of g:

x	0		$\sqrt{3} - 1$		$+\infty$
g'		$+$	0	$-$	
g	$\frac{2}{5}$	$\nearrow$	$\frac{5\sqrt{3}}{18}$	$\searrow$	0

On $[1, 2]$, the function g is decreasing, so, $\forall x \in [1, 2]$,

$$g(x) \leq g(1) \simeq 0.47 = K < 1.$$

Therefore,

$$|F'(x)| \leq 0.48 \simeq K < 1 \quad K = \frac{80}{169},$$

which means that F is a contraction mapping. This implies that the fixed-point algorithm converges, and $\lim_{n \to +\infty} x_n = \alpha$ for $x_0 = 1$, where $\alpha = F(\alpha)$.

Note that:

$$|\alpha - x_{24}| = |F(\alpha) - F(x_{23}|.$$

Now, let

$$e_{24} = \alpha - x_{24} = F(\alpha) - F(x_{23}) = F'(\alpha - x_{23}).$$

Then:

$$|e_{24}| \leq \frac{80}{169}|e_{23}| \leq \cdots \leq \left(\frac{80}{169}\right)^{24} |\alpha - x_0| < \left(\frac{80}{169}\right)^{24}.$$

To finish, observe that

$$\left(\frac{80}{169}\right)^{24} \simeq 1.6 \times 10^{-8}.$$

We can now apply the fixed-point method to the function $f(x) = x^3 + 2x^2 + 10x - 20$.

We have the values $y_0 = 1$, $c = 2$, $f(\frac{3}{2}) = \frac{23}{8}$ and $f(2) = 16$, so

$$y_{n+1} = y_n - \frac{f(y_n)}{f(y_n) - f(c)}(y_n - c).$$

Setting

$$\begin{aligned} G(y) &= y - \frac{f(y)}{f(y) - f(c)}(y - c) \\ &= \frac{-yf(c) + cf(y)}{f(y) - f(c)} \end{aligned}$$

gives

$$G'(y) = \frac{[-f(c) + cf'(y)][f(y) - f(c)] - [-yf(c) + cf(y)]f'(y)}{(f(y) - f(c))^2}$$
$$= \frac{-f(c)f(y) + f(c)yf'(y) - cf(c)f'(y) + f(-c)^2}{(f(y) - f(c))^2}.$$

Replacing $f(y)$ with its formula, we find

$$G'(y) = \frac{32(y^3 - 2y^2 - 4y + 8)}{(f(y) - 16)^2}.$$

It can be shown that $|G'(y)| \leq K < 1$ in the neighborhood of α. This implies that

$$|G'(y)| \leq \frac{32|y^3 - y^2 - 4y + 8|}{(f(\frac{3}{2} - 16)^2} < \frac{32.3^3}{13.125^2} \simeq 0.557.$$

The following computations are performed with $c = \frac{3}{2}$. In this case,

$$y_1 = G(y_0) = 1 + \frac{28}{79} = \frac{107}{79} \simeq 1.354430797$$

and $f(y_1) \simeq -0.302055212$;

$$y_2 = G(y_1) = y_1 - \frac{f(y_1)}{f(y_1) - \frac{23}{8}}(\frac{107}{79} - \frac{3}{2})$$
$$\simeq 1.36827067688$$

and $f(y_2) \simeq -0.011685720$;

$$y_3 = G(y_2) = y_2 - \frac{f(y_2)}{f(y_2) - \frac{23}{8}}(y_2 - \frac{3}{2})$$
$$\simeq 1.36878803936$$

and $f(y_3) \simeq -0.000088007$;

$$y_4 = G(y_3) = y_3 - \frac{f(y_3)}{f(y_3) - \frac{23}{8}}(y_3 - \frac{3}{2})$$
$$\simeq 1.3688079520.$$

Therefore,

$$\begin{aligned}|\alpha - y_n| &\leq |\alpha - x_{24}| + |x_{24} - y_4| \\ &\leq 1.6 \times 10^{-8} + 2.5 \times 10^{-7} \\ &\leq 2.2 \times 10^{-7}.\end{aligned}$$

With the secant method, an order of $n = 4$ leads to an accuracy of the order of 10^{-7}, which is much better than the fixed-point method, which needed an order of $n = 24$ to achieve an accuracy of the order of 10^{-8}. Thus, the secant method converges more quickly than the fixed-point method.

Newton's method can be stated as

$$x_{n+1} = x_n - \frac{f(x_n)}{f'(x_n)} = F(x_n),$$

where $f(x) = x^3 + 2x^2 + 10x - 20$ and $f'(x) = 3x^2 + 4x + 10$, which gives

$$x_{n+1} = x_n - \frac{x_n^3 + 2x_n^2 + 10x_n - 20}{3x_n^2 + 4x_n + 10}.$$

Choosing $x_0 = 1$, we find:

$$\begin{aligned}x_1 &\simeq 1.411764706 \\ x_2 &\simeq 1.369336471 \\ x_3 &\simeq 1.368808189 \\ x_4 &\simeq 1.368898108.\end{aligned}$$

In this example, we almost achieved the same accuracy as Leonardo in just four iterations, with an error of 10^{-8}.

Newton's method is significantly better than the secant method, but in a sense they are similar. We can show that the error of Newton's method decays quadratically.

From the second-order Taylor expansion of f about α, we can write that:

$$f(\alpha) = f(x_{n-1}) + (\alpha - x_{n-1})f'(x_{n-1}) + \frac{1}{2}(\alpha - x_{n-1})^2 f''(\xi).$$

Newton's formula gives

$$0 = f(x_{n-1}) + (x_n - x_{n-1})f'(x_{n-1})$$

which implies that

$$f(\alpha) = 0 = (\alpha - x_n)f'(x_{n-1}) + \frac{1}{2}(\alpha - x_{n-1})^2 f''(\xi).$$

By setting $e_n = \alpha - x_n$, we find that

$$0 = e_n f'(x_{n-1}) + \frac{1}{2}e_{n-1}^2 f''(\xi),$$

hence

$$|e_n| \leq \frac{1}{2}\frac{M}{m}|e_{n-1}|^2,$$

where $M = \sup_{[1,\frac{3}{2}]} |f''(x)|$ and $m = \inf_{[1,\frac{3}{2}]} |f'(x)|$. We deduce that

$$|e_n| \leq \left(\frac{M}{2m}\right)^{2^n - 1} |e_0|^{2^n}.$$

1.3.8. Regula falsi *method (or false position method)*

If we approximate $F'(x)$ by the expression:

$$\frac{F(x_n) - F(x_0)}{x_n - x_0}, \qquad [1.5]$$

then the recurrence formula becomes

$$x_{n+1} = x_n - F(x_n)\frac{x_n - x_0}{F(x_n) - F(x_0)}, \quad x_0, x_1 \in [a, b]. \qquad [1.6]$$

This time, we are replacing the tangent to the curve of $F(x)$ at the point $(x_n, F(x_n))$ with the straight line that passes through this point and $(x_0, F(x_0))$.

EXAMPLE.– Consider the function

$$f(x) = e^{-2x} - \cos(x) - 3.$$

The function f is decreasing on $[-1, 0]$, $f(-1) = 3.848754$ and $f(0) = -3$, hence f has a unique zero in $[-1, 0]$.

By applying the *regula falsi* method to f on $[-1, 0]$, we find:

i	x_i	i	x_i
1	−0.438036	8	−0.6656762
2	−0.595945	9	−0.665706
3	−0.645201	10	−0.665714
4	−0.659764	11	−0.665717
5	−0.663996	12	−0.665717
6	−0.665221	13	−0.665718
7	−0.665574	14	−0.665718

An approximate value for the root of f is given by $x_{13} = -0.665718$.

By applying Newton's method to f with initial point $x_0 = 0$, we find that

i	x_i
1	−1.500000
2	−1.086704
3	−0.798386
4	−0.681373
5	−0.665953
6	−0.665718

This also leads to the approximation $x_6 = -0.665718$.

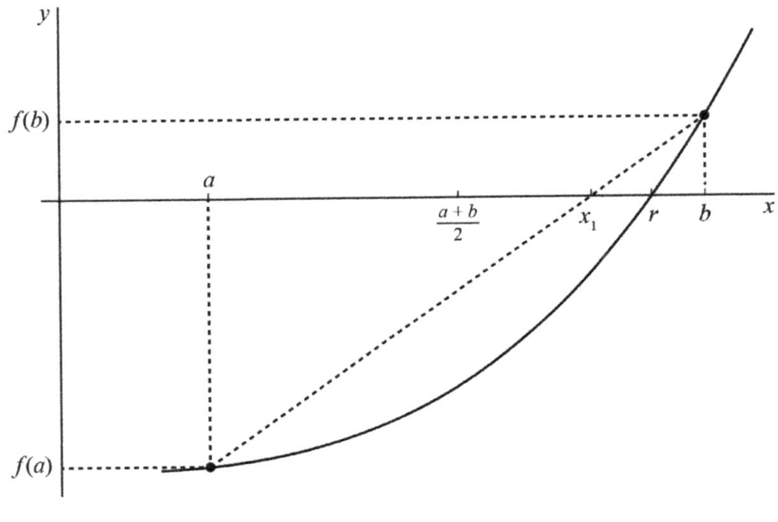

Figure 1.3. Regula falsi *method*

1.4. Order of an iterative process

DEFINITION.– *Let $\bar{x}$ be a root of the equation $F(x) = 0$, and let (x_n) be a numerically constructed sequence that converges to $\bar{x}$.*

We say that the method associated with this sequence is of order k if the following condition holds:

$$x_{n+1} - \bar{x} = (x_n - \bar{x})^k [a_k + \epsilon(x_n)],$$

where a_k is a non-zero constant and $\epsilon(x)$ is a function that satisfies $\lim \epsilon(x) = 0$ *(i.e. $\lim_{n \to +\infty} \epsilon(x_n) = 0$). In other words:*

$$\lim_{n \to =\infty} \frac{x_{n+1} - \overline{x}}{(x_n - \overline{x})^k} = a_k.$$

REMARK.–

1) In practice, the sequence (x_n) is not explicitly known. We say that the method is of at least order k if we can find a constant $M > 0$ and a given rank n_0, such that, for $n \geq n_0$, $|x_{n+1} - \bar{x}| \leq M|x_n - \bar{x}|^k$.

If so, an upper bound for the error of the method is given by:

$$|x_n - \bar{x}| \leq \frac{1}{\sqrt[k-1]{M}} (\sqrt[k-1]{M} |x_0 - \bar{x}|)^{k^n}.$$

2) If the iterative method is defined using a function with continuity class C^p, the order k can be defined as the smallest integer m, such that $f^{(m)}(\bar{x}) \neq 0$ and $a_k = \dfrac{f^{(k)}(\bar{x})}{k!}$ (or alternatively, $M = \sup \left| \dfrac{f^{(k)}(x)}{k!} \right|$).

3) Higher order iterative methods converge faster.

1.5. Using *Matlab*

1.5.1. *Finding the roots of polynomials*

Suppose that we wish to compute the roots of the equation:

$$x^5 - 2x^4 + 2x^3 + 3x^2 + x + 4 = 0.$$

We can use the *roots* function to do this in *Matlab*. First, we enter the coefficients of the polynomial:

```
>> c= [1 -2 2 3 1 4];
```

Next, we can find the solution by running the following command:

```
>> solution = roots(c)

solution =
          1.5336 + 1.4377i
          1.5336 - 1.4377i
         -1.0638
         -0.0017 + 0.9225i
         -0.0017 - 0.9225i
```

Given an array of roots, the function *poly* returns the coefficients of the polynomial with these roots ordered by decreasing powers. Thus, in this example, we can recover the original polynomial as follows:

```
>> poly(solution)

ans =
     1.0000 -2.0000 2.0000 3.0000 1.0000 4.0000
```

Matlab allows us to find the zero of a function using the *fzero* command. To do this, we need to create a file (e.g. 'f.m') containing the function. In this example, we shall choose $f(x) = \exp(x) - 2\cos(x)$.

```
function f=f(x)
f=exp(x)-2*cos(x)
```

To find the solution of $f(x) = 0$ in the neighborhood of the point $x = 0.5$, we run the command 'fzero('f',0.5)'. This function displays the values of $f(x)$ computed at each iteration until the desired solution x^* is found.

```
>>d=fzero('f',0.5)

f =
     -0.1064
     -0.1430
     -0.0692
     -0.1579
```

```
      -0.0536
      -0.1788
      -0.0313
      -0.2080
       5.8950e-004
      -3.0774e-005
      -4.1340e-009
       2.2204e-016
      -8.8818e-016

d =
      0.5398
```

1.5.2. *Bisection method*

To implement the bisection method, we can create a script called 'bisection.m', with the following code:

```
a=0;
fa=-5;
b=3;
fb=16;
eps=1.0e-3;
while (b-a) >eps
  x=(a+b)/2;
  fx=x^3-2*x-5;
  if (sign(fx) == sign(fa))
     a=x;
     fa=fx;
  else
     b=x;
     fb=fx;
  end
end;
x
```

We can run the script 'bisection.m' as follows to compute the root of the function $f(x) = x^3 - 2x - 5$ contained in the interval $[0, 3]$:

```
>> bisection

x=
   2.0940
```

1.5.3. *Newton's method*

The following script, saved under the name 'newton.m', implements Newton's method:

```
format
short;
clc;
syms x;

display ('Newton''s method is an iterative method for solving
            nonlinear systems');
f = input('Enter the function f(x)=');

x0 = input('Enter the initial estimate of the solution: ');

n = input('Enter the maximum number of iterations: ');

e = input('Enter the margin of error: ');
i = 0;
while (i<n)
    df = diff(f,x);
    dfx0 = subs(df,x,x0);
    fx0 = subs(f,x,x0);
    x1 = x0 - (fx0/dfx0);
    dx = x1 - x0;
   if abs(dx)/abs(x0)<=e
      fprintf('The solution is = %i\n',x1);
      fprintf('The number of iterations required was = %i\n',i);
      break
   end
    i = i+1;
    x0=x1;
end
if abs(dx)/abs(x0) >e
    fprintf('Convergence was not achieved in n iterations ');
end
```

We can apply this script to compute the root of the equation $x^3 - 3x + 1 = 0$. Executing the script 'newton.m' returns the following output:

```
Newton's method is an iterative method for solving nonlinear systems
Enter the function f(x)=x^3-3*x+1
```

```
Enter the initial estimate of the solution: 0

Enter the maximum number of iterations: 100

Enter the margin of error: 0.001

The solution is = 3.472964e-001

The number of iterations required was = 2
```

Therefore, the solution computed by this script is $x = 0.3473$.

2

Numerically Solving Differential Equations

2.1. Introduction

Differential equations are found in a wide variety of fields, including physics (e.g. mechanical physics), chemistry (e.g. reaction kinetics), biology (e.g. population dynamics) and much more.

The simplest example of a demographic model considers an isolated population. At any time t, we denote $P(t)$ the current number of individuals in the population. We shall assume that this number is sufficiently high that we can view $P(t)$ as a real number (rather than as an integer). We can describe the evolution of this population with a differential equation:

$$\frac{dP(t)}{dt} = \text{births - deaths + migration}. \qquad [2.1]$$

If we assume that there is no migration, and that the births and deaths are proportional to the size of the population, we obtain a linear equation:

$$\frac{dP(t)}{dt} = aP(t) - bP(t). \qquad [2.2]$$

This expression for the population increases or decreases exponentially, which is not physically realistic, especially in the long run. We need to add a corrective term that depends on the size of the population to reflect the assumption that an ideal size exists (the biotic capacity). Below this capacity, the population should increase; above this capacity, it should decrease. One example of a model that meets these criteria is known as the logistic model. It has the following expression:

$$\frac{dP(t)}{dt} = rP(t)\left(1 - \frac{P(t)}{K}\right). \qquad [2.3]$$

The constant K denotes the biotic capacity. This equation has the family of solutions:

$$P(t) = \frac{N(0)K\exp(rt)}{K + N(0)(\exp(rt) - 1)}. \quad [2.4]$$

There are two equilibrium points, 0 and K. If $N(0) = 0$, the population size remains permanently fixed at zero. If $N(0) = K$, the population size is also stable, equal to K. Furthermore, if $N(0) > 0$, then it's easy to see that $\lim_{t \to +\infty} N(t) = K$. This means that K is a stable equilibrium point, whereas 0 is unstable: if $N(0)$ is close to K, the solution remains close to K; if $N(0)$ is close to 0, it moves away from zero over time.

We could also model interactions with other populations. As a simple example, consider two populations with a predator–prey relationship. The Lotka–Volterra model proposes one way of describing this dynamic. It makes the following assumptions:

– in the absence of predators, the population of prey should increase exponentially;

– in the absence of prey, the predator death rate should be proportional to the predator population size;

– the rate of predation should be proportional to the number of meetings between predator and prey, which is assumed to be proportional to the product of the two population sizes;

– the predator growth rate should also be proportional to the number of meetings between the predator and the prey.

Writing N for the prey population and P for the predator population, we obtain the following system of equations:

$$\frac{dN(t)}{dt} = N(a - bP), \quad [2.5]$$

$$\frac{dP(t)}{dt} = P(cN - d). \quad [2.6]$$

The four constants a, b, c and d are positive. This is a dynamic system with very specific properties. It has two equilibrium points, $(0, 0)$ and $(\frac{d}{c}, \frac{a}{b})$. In the phase space, that is, the (planar) space spanned by the coordinates N and P, the system follows a closed path, implying that N and P are periodic. This model is not perfect. For instance, in the absence of predators, it is unrealistic to assume that the prey population will increase indefinitely.

These examples have allowed us to explore a number of different scenarios: linear equations, nonlinear equations and systems of equations. However, we had no reason to believe *a priori* that these equations had solutions, nor did we know whether the stated solutions were unique. The questions of existence and uniqueness are studied in more depth in section 2.

2.2. Cauchy problem and discretization

Suppose that f: $[a,b]\times\mathbb{R}\to\mathbb{R}$ and $\eta\in\mathbb{R}$. We will consider the following Cauchy problem:

Does there exist a function y: $[a,b]\to\mathbb{R}$ that is differentiable on $[a,b]$ and which satisfies [CRO 84]:

$$(\mathcal{P})\begin{cases} y(a)=\eta \\ y'(x)=f(x,y(x))\ \forall x\in[a,b] \end{cases} \qquad [2.7]$$

DEFINITION.– *We say that the function* f: $[a,b]\times\mathbb{R}\to\mathbb{R}$ *is Lipschitz in* y *on* $[a,b]\times\mathbb{R}$, *if there exists* $K>0$, *such that:*

$$|f(x,y)-f(x,z)|\leq K|y-z|,\quad \forall x\in[a,b],\ \forall y,z\in\mathbb{R}.$$

The constant K *is known as the Lipschitz constant.*

THEOREM.– *Consider the Cauchy problem* $(\mathcal{P})$. *If* f: $[a,b]\times\mathbb{R}\to\mathbb{R}$ *satisfies the hypotheses:*

i) f *is continuous;*

i) $|f(x,y)-f(x,z)|\leq K|y-z|\ \forall x\in[a,b],\ \forall y,z\in\mathbb{R}$.

Then the problem $(\mathcal{P})$ *has a unique solution.*

Suppose that the interval $[a,b]$ is partitioned by a sequence of points $(x_i)_{0\leq i\leq N}$ $x_0=a<x_1<x_2<\cdots<x_N=b\quad(x_i=a+ih,\ h=\frac{b-a}{N})$.

Our goal is to find N real numbers $y_1,y_2,\ldots,y_N$, such that each y_i approximates $y(x_i)$. We will then connect these points together by interpolation to define a function y_h on $[a,b]$.

Finally, we will establish an estimate of the discretization error $e_n=y_n-y(x_n)$, which depends on h.

A numerical method is said to be of order p if it yields approximate values y_n of $y(x_n)$ that satisfy $|e_n|\leq Kh^p$.

The two numerical methods presented below calculate an approximate solution y_{i+1} at the point x_{i+1} from the approximate solution at x_i.

Suppose that we wish to solve the differential equation $y'(t) = f(t, y(t))$, where $t \in [0,1]$ and $y(0) = y_0$. We shall assume that f has continuity class C^∞ and is K-Lipschitz. We can use the following method:

$$y_{n+1} = y_n + h(\frac{1}{3}f(t_n, y_n) + \frac{2}{3}f(t_n + \frac{3}{4}h, y_n + \frac{3}{4}hf(t_n, y_n))), \qquad [2.8]$$

where $h = \frac{1}{N}$ and $0 \leq n \leq N-1$. The numerical scheme defined by [2.8] is both consistent and stable. To see this, observe first that it is defined by the relation $y_{n+1} = y_n + hF(t_n, y_n, h)$, where

$$F(t, y, h) = \frac{1}{3}f(t, y) + \frac{2}{3}f(t + \frac{3}{4}h, y + \frac{3}{4}hf(t, y)).$$

Since $F(t, y, 0) = f(t, y)$, this method is consistent. Moreover, since f is L-Lipschitz, the following relation holds $\forall y, z \in [0,1]$:

$$|F(t, y, h) - F(t, z, h)| \leq |y - z|(L + h\frac{L^2}{2}).$$

This shows that the method is stable. As it is both stable and consistent, it converges. It is at least a second-order, which can be shown by computing $\frac{\partial}{\partial h}F(t, y, h)$:

$$\begin{aligned}\frac{\partial}{\partial h}F(t, y, h) &= \frac{2}{3}\left(\frac{3}{4}\frac{\partial}{\partial t}f(t + \frac{3}{4}h, y + \frac{3}{4}h(f(t, y))\right.\\ &\left.+ \frac{3}{4}f(t, y)\frac{\partial}{\partial y}f\left(t + \frac{3}{4}h, y + \frac{3}{4}h(f(t, y))\right)\right)\\ &= \frac{1}{2}\left(\frac{\partial}{\partial t}f(t + \frac{3}{4}h, y + \frac{3}{4}h(f(t, y))\right.\\ &\left.+ f(t, y)\frac{\partial}{\partial y}f\left(t + \frac{3}{4}h, y + \frac{3}{4}h(f(t, y))\right)\right).\end{aligned}$$

Hence,

$$\frac{\partial}{\partial h}F(t, y, 0) = \frac{1}{2}\left(\frac{\partial}{\partial t}f(t, y) + f(t, y)\frac{\partial}{\partial y}f(t, y)\right) = \frac{1}{2}f^{[1]}(t, y).$$

As $F(t, y, 0) = f(t, y)$, this proves that the method is at least of the first-order.

For $f(t, y) = -\lambda y$,

$$F(t, y, h) = \frac{-\lambda y}{3} + \frac{2}{3}(-\lambda)(y + \frac{3}{4}h(-\lambda y)) = -\lambda y + \frac{1}{2}\lambda^2 + hy.$$

Thus,

$$
\begin{aligned}
F(t, y, 0) &= -\lambda y = f(t, y) \\
\frac{\partial}{\partial h} F(t, y, h) &= \frac{1}{2} \lambda^2 y = \frac{\partial}{\partial h} F(t, y, 0) \\
f^{[1]}(t, y) &= 0 + (-\lambda y)(-\lambda) = \lambda^2 y
\end{aligned}
$$

Therefore, $\frac{\partial}{\partial h} F(t, y, h) = \frac{1}{2} f^{[1]}(t, y)$, and so the method is at least of the second-order. Finally,

$$
\frac{\partial^2}{\partial h^2} F(t, y, h) = 0,
$$

but $f^{[2]}(t, y) = (-\lambda y) y^2 = -\lambda^3 y$. This shows that the method is not of the third-order.

Applied to the example problem $y' = -\lambda y$, the method has the expression:

$$
y_{n+1} = y_n + h(-\lambda y_n + \frac{1}{2} \lambda^2 h y_n) = y_n (1 - h\lambda + \frac{(h\lambda)^2}{2}) = y_n P(h\lambda),
$$

where $P(x) = 1 - x + \frac{x^2}{2} = \frac{1}{2}((x - 1)^2 + 1)$ (this is identical to the augmented Euler method, with the same radius of stability). This method convergences whenever $|P(h\lambda)| < 1$. Its radius of stability is equal to 2.

EXAMPLE.– Consider the Cauchy problem:

$$
\begin{cases} y'(t) = t^2 - y(t) \\ y(0) = \quad 1 \end{cases} \qquad [2.9]
$$

It is easy to verify that the analytic solution of [2.9] is $y(t) = -e^{-t} + t^2 - 2t + 2$.

Now, consider the following method:

$$
y_{n+1} = y_n + \frac{3}{4} h f(t_{n+1}, y_{n+1}) + \frac{1}{4} h f(t_n, y_n). \qquad [2.10]
$$

We can use this method to approximate problem (1). Here, $f(t, y(t)) = t^2 - y(t)$ and h is the time step.

For $h = 0.2$, after two iterations, we find: $y_1 = 0.8313$ and $y_2 = 0.7093$.

For $h = 0.1$, after four iterations, we find: $y_1 = 0.9077$, $y_2 = 0.8263$, $y_3 = 0.7566$ and $y_4 = 0.6995$. Note that $y_4 = 0.6995$ is the closest to the exact value, which is $y(0.4) = 0.6897$.

2.3. Euler's method

Euler's method can be written as follows:

$$\begin{cases} y_{i+1} = y_i + hf(x_i, y_i) \\ y_0 = y(a) = \eta \\ 1 \leq i \leq N \end{cases} \qquad [2.11]$$

2.3.1. *Interpretation*

Euler's method is based on the assumption that, on the interval $[x_0, x_0 + h]$, the curve is approximately equal to its tangent at x_0, which has equation $z(x) = y'(x_0)(x - x_0) + y(x_0)$. The formula $y_1 = y_0 + hf(x_0, y_0)$ therefore gives a good approximation of $y(x_1)$.

Next, we assume that $f(x_1, y_1)$ is a good approximation of $y'(x_1)$ and, on the interval $[x_1, x_2]$, we replace the curve by its approximate tangent at x_1, which has equation: $z(x) = y'(x_1)(x - x_1) + y(x_1) = f(x_1, y_1)(x - x_1) + y_1$.

Therefore, we can use $y_2 = y_1 + hf(x_1, y_1)$ to approximate $y(x_2)$.

2.3.2. *Convergence*

THEOREM.– *If f satisfies the following hypotheses:*

i) $f \in C^1([a, b] \times \mathbb{R})$;

ii) f is K-Lipschitz in y:

$$|f(x, y) - f(x, z)| \leq K|y - z| \quad \forall x \in [a, b], \qquad \forall y, z \in \mathbb{R}, \ K > 0;$$

then Euler's method converges.

More precisely, choosing $M = \max\{|y'(t)|, t \in [a, b]\}$, the following expression gives an upper bound for the error:

$$|e_n| = |y_n - y(x_n)| \leq \frac{e^{K(b-a)} - 1}{K} \frac{M}{2} h$$

and

$$\lim_{n\to\infty} \max_{n=1,\cdots,N} |e_n| = 0.$$

REMARK.–

1) The result of this theorem can be summarized as $|e_n| \le Ah$, where $A > 0$ is a constant; in other words, Euler's method is of the first-order.

2) First-order methods do not converge sufficiently quickly to give results that are useful in practice. We will present a higher-order numerical method below.

EXAMPLE.– Consider the differential equation:

$$\begin{cases} y' &= 2y \\ y(0) &= 5 \end{cases} \qquad (1)$$

The analytic solution is $y(t) = 5e^{2t}$. Setting $h = \frac{1}{n}$ for $i = 0, \ldots, n$, Euler's method gives the approximations $y_i = 5(1+2h)^i$. For $i = 0, \ldots, 100$, we define $t_i = ih$ $(h = 0.01)$. We can now compute the values $y(t_i)$, y_i for $i = 0, 10, 20, \ldots, 100$.

t_i	$y(t_i)$	y_i
0	5.	5.
0.1	6.107013790	6.094972100
0.2	7.459123490	7.429736980
0.3	9.110594000	9.056807920
0.4	11.12770464	11.04019832
0.5	13.59140914	13.45794016
0.6	16.60058462	16.40515397
0.7	20.27599984	19.99779113
0.8	24.76516212	24.37719581
0.9	30.24823732	29.71566566
1.0	36.94528050	36.22323061

2.4. One-step Runge–Kutta method

Runge–Kutta methods are widely used in practice because they have various advantages (easy to program, stable solutions, easy to adjust the step, knowledge of y_0 is sufficient to integrate the differential equation). However, these methods are computationally slow, and it can be difficult to estimate the local error.

2.4.1. *Second-order Runge–Kutta method*

This method considers the first-order centered differences:

$$x_{i1} = x_i + \theta_1, h, \ \ 0 \le \theta_1 \le 1$$

$$y_{i1} = y_i + hA_{10} f(x_i, y_i)$$

$$y_{i+1} = y_i + h[A_{20} f(x_i, y_i) + A_{21} f(x_{i1}, y_{i1})].$$

The values of the coefficients θ_1, A_{10}, A_{20}, A_{21} must be chosen in such a way that the method is of the second-order.

Writing $f_i = f(x_i, y_i)$,

$$\begin{aligned} y_{i+1} &= y_i + h\left[A_{20} f_i + A_{21} f(x_i + \theta_1 h, y_i + hA_{10} f_i)\right] \\ &= y_i + h\left[A_{20} f_i + A_{21} f_i + A_{21}\theta_1 h f'_{xi} + A_{21} A_{10} f_i h f'_{yi} + ...\right] \end{aligned}$$

$$y(t_{i+1}) = y(x_i) + hy'(x_i) + \frac{h^2}{2} y''(x_i) + \frac{h^3}{6} y^{(3)}(\zeta_i)$$

$$e_{i+1} = y_{i+1} - y(x_{i+1}).$$

This gives the system

$$\begin{cases} A_{20} + A_{21} = 1 \\ A_{21} A_{10} = \frac{1}{2} \\ A_{21}\theta_1 = \frac{1}{2}. \end{cases} \qquad [2.12]$$

This is a nonlinear system of 3 equations and 4 unknowns. The usual approach is to fix θ_1:

$$A_{21} = \frac{1}{2\theta_1} \qquad [2.13]$$

$$A_{20} = \frac{2\theta_1 - 1}{2\theta_1} \qquad [2.14]$$

$$A_{10} = \theta_1. \qquad [2.15]$$

EXAMPLE.– Consider the differential equation:

$$\begin{cases} y' &= -y + t + 1 \\ y(0) &= 1 \end{cases} \qquad (1)$$

The analytic solution is $y(t) = t + e^{-t}$. We write y_i and z_i for the values found using the second-order Taylor method and the second-order Runge–Kutta method, respectively.

t_i	$y(t_i)$	y_i	z_i
0	1.	1.	1.
0.1	1.004837418	1.005000000	1.005000000
0.2	1.018730753	1.019025000	1.019025000
0.3	1.040818221	1.041217625	1.041217625
0.4	1.070320046	1.070801950	1.070801951
0.5	1.106530660	1.107075765	1.107075766
0.6	1.148811636	1.149403567	1.149403568
0.7	1.196585304	1.197210228	1.197210229
0.8	1.249328964	1.249975256	1.249975257
0.9	1.306569660	1.307227606	1.307227608
1.0	1.367879441	1.368540983	1.368540985

2.4.2. *Fourth-order Runge–Kutta method*

The most common version of Runge–Kutta is the fourth-order method (RK4), which can be used whenever the functions being studied are sufficiently regular:

$$x_{ij} = x_i + \theta_j h \tag{2.16}$$

$$y_{ij} = y_i + h \sum_{k=0}^{j-1} A_{jk} f(x_{ik}, y_{ik}) \tag{2.17}$$

$$A_{10} = \theta_1 \tag{2.18}$$

with $j = 1, 2, 3, 4$ and $y_{i3} = y_{i+1}$.

The coefficients θ_j are chosen in such a way that the method is of the fourth-order. This leads to:

$$\theta_1 = \frac{1}{2},\ \theta_2 = \frac{1}{2},\ \theta_3 = \theta_4 = 1. \tag{2.19}$$

$$A_{10} = \frac{1}{2} \tag{2.20}$$

$$A_{20} = 0, \quad A_{21} = \frac{1}{2} \tag{2.21}$$

$$A_{30} = 0, \quad A_{31} = 0, \quad A_{32} = 1 \tag{2.22}$$

$$A_{40} = \frac{1}{6}, \quad A_{41} = \frac{1}{3},\ A_{42} = \frac{1}{3},\ A_{43} = \frac{1}{6}. \tag{2.23}$$

Therefore, we can compute y_{i+1} as follows:

$$y_{i+1} = y_i + \frac{\Delta t}{6}(f(x_i, y_i) + 2f(x_{i+1/2}, \hat{y}_{i+1/2}) + 2f(x_{i+1/2}, \tilde{y}_{i+1/2}) + f(x_{i+1}, \hat{y}_{i+1})) \quad [2.24]$$

$$\hat{y}_{i+1/2} = y_i + \frac{\Delta x}{2} f(x_i, y_i) \quad [2.25]$$

$$\tilde{y}_{i+1/2} = y_i + \frac{\Delta x}{2} f(x_{i+1/2}, \hat{y}_{i+1/2}) \quad [2.26]$$

$$\hat{y}_{i+1} = y_i + \Delta t f(x_{i+1/2}, \tilde{y}_{i+1/2}). \quad [2.27]$$

To compute y_{i+1}, we need to evaluate the function f four times. This can lead to high computation times with complicated functions.

EXAMPLE.– Consider the differential equation:

$$\begin{cases} y' &= ty \\ y(1) &= 3 \end{cases} \quad (1)$$

The analytic solution is $y(t) = 3e^{\frac{(t^2-1)}{2}}$. For $i = 0, \ldots, 10$, we define $t_i = 1 + ih$ (with $h = 0.1$). The approximate values of $y(t_i)$ for $i = 0, \ldots, 10$ calculated using Runge–Kutta 4 are shown in Table 2.1.

t_i	$y(t_i)$	y_i
1.0	3.	3.
1.1	3.332131830	3.332131472
1.2	3.738230193	3.738229173
1.3	4.235969760	4.235967573
1.4	4.848223206	4.848219004
1.5	5.604737871	5.604730254
1.6	6.544416795	6.544403471
1.7	7.718440137	7.718417376
1.8	9.194562609	9.194524378
1.9	11.06306728	11.06300382
2.0	13.44506721	13.44496278

Table 2.1. *Approximate values computed by RK4*

EXAMPLE.– Suppose that we wish to find an algorithm to compute the approximate values for the solution y of the problem:

$$\begin{cases} y'(t) = g(t)y \quad t \in [a, b] \\ y(a) = \alpha \end{cases}$$

(where g is a function that we can compute using some procedure $G(S)$) with fourth-order Runge–Kutta).

In this case, $F(t,y,h) = \frac{1}{6}(k_1 + 2k_2 + 2k_3 + k_4)$, where $f(t,y) = g(t)y$, and

$$\begin{aligned} k_1 &= g(t)y \\ k_2 &= g(t+\tfrac{h}{2})(y+\tfrac{h}{2}k_1) \\ k_3 &= g(t+\tfrac{h}{2})(y+\tfrac{h}{2}k_2) \\ k_4 &= g(t+h)(y+hk_3). \end{aligned}$$

We will use the following variables: N is the number of iterations; t is the time variable; a, b are the end points of the interval; y is an array of size $N+1$; k_1, k_2, k_3 and k_4 are computation variables and $G(t)$ is the procedure that computes $g(s)$.

Start

Initialization: a, b, N, z

set size of y to $N+1$

$y(0) := z$

$h := (b-a)/2/N$

$t := a$

For $i = 1$ to N, do

$p := G(t)$

$k1 := z * p$

$t := t + h$

$p := G(t)$

$k2 := (z + h * k1) * p$

$k3 := (z + h * k2) * p$

$t := t + h$

$p := G(t)$

$k4 := (z + 2 * h * k3) * p$

$z := z + 2 * h * (k1 + 2 * k2 + 2 * k3 + k4)/6$

$y(i) := z$

Print t and $y(i)$

End of For

End

2.5. Multi-step Adams methods

Adams methods are a category of multi-step methods. They can be divided into two subcategories: open-formula methods and closed-formula methods.

2.5.1. *Open Adams methods*

Consider the following differential equation:

$$\begin{cases} \dfrac{dy}{dt} = f(y,t) \\ y(0) = y_0 \end{cases} \quad [2.28]$$

The Taylor expansion about t is:

$$y(t+\Delta t) = y(t) + \Delta t\frac{dy}{dt} + \frac{(\Delta t)^2}{2!}\frac{d^2y}{dt^2} + \frac{(\Delta t)^3}{3!}\frac{d^3y}{dt^3} + \frac{(\Delta t)^4}{4!}\frac{d^4y}{dt^4} + \cdots + \frac{(\Delta t)^n}{n!}\frac{d^ny}{dt^n}.$$

In the differential equation [2.28], we can write:

$$\begin{cases} \dfrac{dy}{dt} = y'(t) = f(y,t) \\ \dfrac{d^2y}{dt^2} = y''(t) = f'(y,t) \\ \vdots \\ \dfrac{d^ny}{dt^n} = y^{(n)}(t) = f^{(n-1)}(y,t) \end{cases}$$

where

$$y_{i+1} = y_i + \Delta t f_i + \frac{(\Delta t)^2}{2!}f_i' + \frac{(\Delta t)^3}{3!}f_i'' + \frac{(\Delta t)^4}{4!}f_i^{(3)} + \cdots + \frac{(\Delta t)^n}{n!}f_i^{(n-1)}. \quad [2.29]$$

2.5.1.1. *First-order Adams formulas*

If we keep only the first two terms (up to first-order) in equation [2.29]:

$$y_{i+1} = y_i + \triangle t f_i, \tag{2.30}$$

then we recover Euler's method.

2.5.1.2. *Second-order Adams formulas*

Suppose now that we keep the first three terms in equation [2.29]:

$$y_{i+1} = y_i + \triangle t f_i + \frac{(\triangle t)^2}{2!} f_i' + O(\triangle t)^3. \tag{2.31}$$

If we substitute the formula left finite differences according to:

$$f_i' = \frac{f_i - f_{i-1}}{\triangle t}, \tag{2.32}$$

then equation [2.31] becomes:

$$y_{i+1} = y_i + \triangle t f_i + \frac{(\triangle t)^2}{2!}\left(\frac{f_i - f_{i-1}}{\triangle t} + O(\triangle t)\right) + O(\triangle t)^3. \tag{2.33}$$

Alternatively:

$$y_{i+1} = y_i + \frac{\triangle t}{2}(3f_i - f_{i-1}) + O(\triangle t)^3. \tag{2.34}$$

Third-order terms are neglected, so this formula is of the second-order. To compute y_{i+1}, we need to know f_i and f_{i-1}. Note that we cannot find y_1, since we do not know y_{-1}. Therefore, to initialize this method, we need some other way of evaluating y_1 (e.g. second-order Runge–Kutta).

2.5.1.3. *Third-order Adams formulas*

Suppose now that we keep the first four terms in equation [2.29]:

$$y_{i+1} = y_i + \triangle t f_i + \frac{(\triangle t)^2}{2!} f_i' + \frac{(\triangle t)^3}{3!} f_i'' + O(\triangle t)^4.$$

We can rewrite f_i' and f_i'' as left differences as follows:

$$f_i' = \frac{f_i - f_{i-1}}{\triangle t} + \frac{\triangle t}{2} f_i'' + O(\triangle t)^2, \tag{2.35}$$

$$f_i'' = \frac{f_i - 2.f_{i-1} + f_{i-2}}{(\triangle t)^2} + O(\triangle t). \tag{2.36}$$

After regrouping the various terms, we find the following expression:

$$y_{i+1} = y_i + \frac{\Delta t}{12}(23f_i - 16f_{i-1} + 5f_{i-2}) + O(\Delta t)^4. \quad [2.37]$$

As before, we will need some other method to compute y_1 and y_2, such as fourth-order Runge–Kutta.

2.5.1.4. *Higher-order Adams formulas*

Similarly, we can use a general formula of order $n+1$ to express y_{i+1} as a function of y_i and $f_i, f_{i-1}, f_{i-2}, \dots, f_{i-n}$:

$$y_{i+1} = y_i + \Delta t \sum_{k=0}^{n} \beta_{nk} . f_{i-k} + O(\Delta t)^{n+2}. \quad [2.38]$$

Table 2.2 lists the values of β_{nk} up to $n = 5$, which results in a sixth-order formula.

k,n	0	1	2	3	4	5	Order of the method
0	1						1
1	$\frac{3}{2}$	$-\frac{1}{2}$					2
2	$\frac{23}{12}$	$-\frac{16}{12}$	$\frac{5}{12}$				3
3	$\frac{55}{24}$	$-\frac{59}{24}$	$\frac{37}{24}$	$-\frac{9}{24}$			4
4	$\frac{1901}{720}$	$-\frac{2774}{720}$	$\frac{2616}{720}$	$-\frac{1274}{720}$	$\frac{251}{720}$		5
5	$\frac{4277}{1440}$	$-\frac{7923}{1440}$	$\frac{9982}{1440}$	$-\frac{7298}{1440}$	$\frac{2877}{1440}$	$-\frac{475}{1440}$	6

Table 2.2. *The values of β_{nk}*

The fourth-order formula is the most widely used:

$$y_{i+1} = y_i + \frac{\Delta t}{24}(55f_i - 59f_{i-1} + 37f_{i-2} - 9f_{i-3}) + O(\Delta t)^5. \quad [2.39]$$

2.5.2. *Closed Adams formulas*

The second class of Adams formulas uses the "backwards" Taylor expansion:

$$\begin{aligned} y(t) &= y(t + \Delta t - \Delta t) \\ &= y(t+\Delta t) - \Delta t y'(t+\Delta t) + \frac{(\Delta t)^2}{2!} y''(t+\Delta t) - \frac{(\Delta t)^3}{3!} y^{(3)}(t+\Delta t) \\ &\quad + \cdots + (-1)^n \frac{(\Delta t)^n}{n!} y^{(n)}(t+\Delta t). \end{aligned} \quad [2.40]$$

In this Taylor expansion, we observe that:

$$\begin{cases} y' &= f \\ y'' &= f' \\ \vdots & \\ y^{(n)} &= f^{(n-1)} \end{cases} \quad [2.41]$$

Hence, for $t = t_i$:

$$\begin{aligned} y_i &= y_{i+1} - \Delta t f_{i+1} + \frac{(\Delta t)^2}{2!} - \frac{(\Delta t)^3}{3!} f''_{i+1} \\ &+ \frac{(\Delta t)^4}{4!} f^{(3)}_{i+1} + \cdots + (-1)^n \frac{(\Delta t)^n}{n!} f^{(n-1)}_{i+1}. \end{aligned} \quad [2.42]$$

This gives:

$$\begin{aligned} y_{i+1} &= y_i + \Delta t f_{i+1} - \frac{(\Delta t)^2}{2!} + \frac{(\Delta t)^3}{3!} f''_{i+1} \\ &- \frac{(\Delta t)^4}{4!} f^{(3)}_{i+1} + \cdots + (-1)^{(n+1)} \frac{(\Delta t)^n}{n!} f^{(n-1)}_{i+1}. \end{aligned} \quad [2.43]$$

This formula is said to be "closed" because we need to compute $f_{i+1} = f(y_{i+1}, t_{i+1})$ to find the unknown y_{i+1}, and f_{i+1} itself usually depends on y_{i+1}. Hence, we need to apply an iterative method.

The idea is to inject an initial estimate $y^{(0)}_{i+1}$ of y_{i+1} into the equation. This gives a new value $y^{(1)}_{i+1}$:

$$y^{(1)}_{i+1} = y_i + \Delta t f(y^{(0)}_{i+1}, t_{i+1}). \quad [2.44]$$

Next, we inject $y_{i+1}^{(1)}$ into the equation to calculate $y_{i+2}^{(2)}$. The calculations are terminated when convergence is attained. For example, if the convergence criterion is defined in terms of an accuracy threshold ϵ, we stop when:

$$|y_{i+1}^{(r+1)} - y_{i+1}^{(r)}| \leq \epsilon. \quad [2.45]$$

At order $n+1$, this leads to the following generalized formula:

$$y_{i+1} = y_i + \Delta t \sum_{k=0}^{n} \gamma_{nk} f_{i+1-k} + O(\Delta t)^{n+2}. \quad [2.46]$$

The values of γ_{nk} are listed up to $n = 5$ in Table 2.3.

k,n	0	1	2	3	4	5	Order of the method
0	1						1
1	$\frac{1}{2}$	$\frac{1}{2}$					2
2	$\frac{5}{12}$	$\frac{8}{12}$	$-\frac{1}{12}$				3
3	$\frac{9}{24}$	$\frac{19}{24}$	$-\frac{5}{24}$	$\frac{1}{24}$			4
4	$\frac{251}{720}$	$\frac{646}{720}$	$-\frac{264}{720}$	$\frac{106}{720}$	$-\frac{19}{720}$		5
5	$\frac{1475}{1440}$	$\frac{1427}{1440}$	$-\frac{798}{1440}$	$\frac{482}{1440}$	$-\frac{173}{1440}$	$\frac{27}{1440}$	6

Table 2.3. *The values of* γ_{nk}

The fourth-order version of this method is again the most widely used:

$$y_{i+1} = y_i + \frac{\Delta t}{24}(9f_{i+1} + 19f_i - 5f_{i-1} + f_{i-2}) + O(\Delta t)^5. \quad [2.47]$$

Closed formulas, which are iterative, require higher computation times than open formulas, since we need to compute $|y_{i+1}^{(r+1)} - y_{i+1}^{(r)}|$ until the difference converges to within ϵ.

REMARK.– The closed version of the method at any given order is much more accurate and stable than the corresponding "open" version. The closer the initial estimate $y_{i+1}^{(0)}$ to the exact value y_{i+1}, the faster the convergence.

2.6. Predictor–Corrector method

The two methods presented above (closed and open Adams formulas) can be combined into a method known as the predictor–corrector method. The idea is to begin by finding an estimated value $y_{i+1}^{(0)}$, the predictor that is close to the final value y_{i+1}. This predictor $y_{i+1}^{(0)}$ is computed using the open version of the formula. The value of $y_{i+1}^{(0)}$ thus obtained is then injected into the closed method of the same order. This combined approach accelerates the convergence, allowing us to fully exploit the advantages of the Adams method: reduced computation times and significantly decreased estimated truncation errors.

The fourth-order predictor–corrector method can be summarized as follows:

PREDICTOR.– Fourth-order open Adams formula:

$$y_{i+1}^{(0)} = y_i + \frac{\Delta t}{24}(55f_i - 59f_{i-1} + 37f_{i-2} - 9f_{i-3}). \quad [2.48]$$

We know $y_i^{(0)}$ from the previous step. This is taken as the value of the predictor:

$$\hat{y}_{i+1}^{(0)} = y_{i+1}^{(0)} - \frac{251}{720}(y_i - y_i^{(0)}). \quad [2.49]$$

CORRECTOR.– Calculated from the closed Adams formula (e.g. of the fourth-order):

$$\begin{cases} y_{i+1}^{(r+1)} &= y_i + \frac{\Delta t}{24}(9f_{i+1}^{(r)} + 19f_i - 5f_{i-1} + f_{i-2}) \\ f_{i+1}^{(0)} &= f(\hat{y}_{i+1}^{(0)}, t_{i+1}) \end{cases} \quad [2.50]$$

The values y_1, y_2, and y_3 can be computed from Runge–Kutta formulas. Since $y_3^{(0)}$ does not exist, the value y_4^0 is calculated from equations [2.48] and [2.49], and $y_4^{(0)}$ is used to initialize the iterative computations (formula [2.50]). We can estimate $y_5^{(0)}$ from y_5 using equation [2.48], and then we can estimate $\hat{y}_5^{(0)}$ using formula [2.49]. This value is then injected into formula [2.50] to begin the iterative computations.

EXAMPLE.– Consider the implicit method defined by:

$$\begin{cases} y_0 &= y(0) \\ y_{n+1} &= y_n + \frac{h}{2}(f(t_n, y_n) + f(t_{n+1}, y_{n+1})) \end{cases}$$

We can design a prediction–correction algorithm for this method by choosing the predictor $y_{n+1} = y_n + hf(t_n, y_n)$. (We are assuming that a procedure $F(S, Z)$ is available to compute the value of f.) Thus:

Start

Initialization: t_{max}, N, z (initial value of y_0), M, and the size of y is $N+1$:

$y(0) := z$

$h := t_{max}/N$

$t := 0$

$p := F(t, z)$

For $i = 1$ to N, do

$q := p$

$s := t + h$

$x \; : \; = z + h * q$ (prediction)

For $k := 1$ to M (correction loop)

$p := F(s, x)$

$x := z + h/2 * (q + p)$

End of For

$z := x$

$y(i) := z$

Print s and $y(i)$

End of For

End

Suppose now that we wish to find the stability radius of this method (without prediction–correction). The implicit method can be written as

$$y_{n+1} = y_n + \frac{h}{2}(-Ay_n - Ay_{n+1}),$$

so

$$y_{n+1} = \frac{1 - \frac{Ah}{2}}{1 + \frac{Ah}{2}} y_n.$$

Since $\left| \frac{1 - \frac{Ah}{2}}{1 + \frac{Ah}{2}} \right| < 1$ for all h, the method is A-stable and its stability radius is infinite.

2.7. Using *Matlab*

The built-in features offered by *Matlab* can solve almost any problem phrased in terms of ordinary differential equations of the following type and with the following initial conditions:

$$y = f(t, y), \quad y(t_0) = y_0.$$

The first step is to create a *Matlab* function describing the differential system in terms of equations. This function should be of the form

```
function dY = odefct(x,Y)
```

where *odefct* is the (arbitrary) name of the *Matlab* function implementing the mathematical function F. It should return a column vector *dY* containing the components F_1 and F_2 of the function *F*.

REMARK.– Even if the function F does not explicitly depend on x, the function must have two input parameters.

Matlab has several differential equation solvers: some designed for classical problems, and others designed for the so-called "stiff" problems. The following solvers are classical solvers:

– *ode45* uses an explicit one-step Runge–Kutta method. This solver is the best choice for most problems;

– *ode23* also uses an explicit one-step Runge–Kutta method. It can be more efficient than *ode45* in some cases;

– *ode113* uses an Adams–Bashforth–Moulton method. This is a multi-step solver.

There are four solvers for "stiff" problems: *ode15s*, *ode23s*, *ode23t* and *ode23tb*.

The following syntax is used to call each of these solvers:

```
[t,Y] = odexx(@odefct,[t0,t1],Y0)
```

where

– *odexx* is the name of the solver, from the choices listed above;

– *odefct* is the name of the *Matlab* function that implements the mathematical function F of the differential system;

– $[t0, t1]$ is the time interval on which the solution should be computed;

– $Y0$ is the column vector containing the initial data: $Y_0 = (y(t_0), y'(t_0))$.

Each solver has the following output parameters:

– t: column vector containing the nodes of the interval $[t_0, t_1]$ at which the solution was calculated;

– Y: matrix containing the values of the solution and its derivatives at the nodes of the interval $[t_0, t_1]$ specified by the vector t. The i-th row of Y contains the values of the $(i-1)$-th derivative at each of the nodes, which are all in $[t_0, t_1]$. The first row contains the solution y of the original differential equation (E).

REMARK.– Note that the symbol @ is required before *odefct* in the function parameters.

Matlab includes a (short) demo for solving ordinary differential equations, which can be run by typing:

```
>> odedemo
```

Suppose, for example, that we wish to solve the differential equation:

$$\begin{cases} \dfrac{dy}{dt} &= -0.1(y-10) \\ y_0 &= 100 \end{cases}$$

First, we implement the equations of this function in a file named 'myode.m':

```
function yprime = myode(t,y)
yprime = -0.1 * (y-10);
```

We can now find the evolution over time of the function y by solving the differential equation with *ode45* as follows:

```
>> tinitial = 0;
>> tfinal = 60;
>> y0 = 100;
>> [t y] = ode45('myode', [tinitial tfinal], y0)
>> plot(t,y)
```

The *ode45* function and the other solvers also work with systems of multiple coupled differential equations.

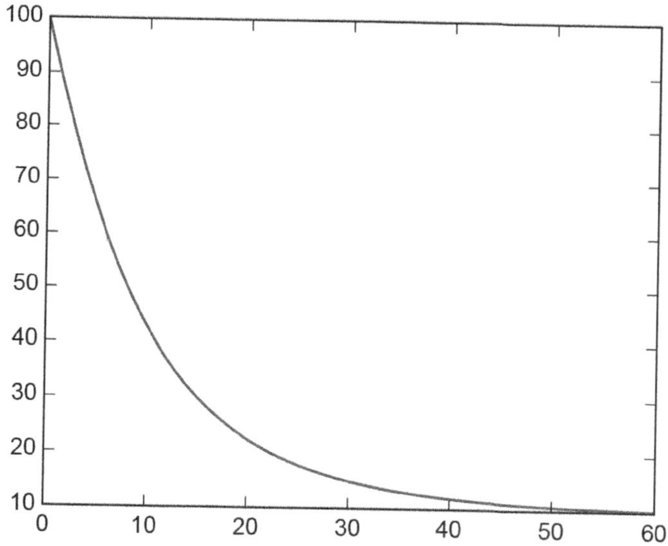

Figure 2.1. *Graphical solution*

PART 2

Solving PDEs

3

Finite Difference Methods

3.1. Introduction

The evolution of many physical problems is described by differential equations with multiple parameters (often t and x), typically involving partial derivatives with respect to each parameter:

– the wave equation:

$$c^2\frac{\partial^2 u}{\partial x^2} - \frac{\partial^2 u}{\partial t^2} = f(x,t); \qquad [3.1]$$

– the (heat) diffusion equation:

$$D\frac{\partial^2 u}{\partial x^2} - \frac{\partial u}{\partial t} = f(x,t); \qquad [3.2]$$

– Schrödinger's equation:

$$\frac{\hbar^2}{2m}\frac{\partial^2 u}{\partial x^2} + i\hbar\frac{\partial u}{\partial t} - U(x)u = 0. \qquad [3.3]$$

These equations are known as Partial Differential Equations (PDEs).

They are often more complex to solve than equations within a single parameter, which are known as Ordinary Differential Equations (ODEs). To solve PDEs, we need to use grids to discretize every parameter simultaneously. For example, a 3D grid is required for space and a 1D grid is required for time (see Figure 3.1).

The best technique to solve the problem varies somewhat as a function of the number of dimensions. We will mostly focus on 2D systems in this chapter, with one spatial dimension and one time dimension, but these methods can also be generalized to N dimensions. Just like ODEs, attempts to solve PDEs encounter obstacles such as:

– precision;

– speed;

– stability.

There are two major frameworks for solving PDEs [CIA 90]:

– Physical space: This class of methods works in "physical" space (real space). This includes methods such as:

- finite difference methods;
- finite element methods;
- finite volume methods.

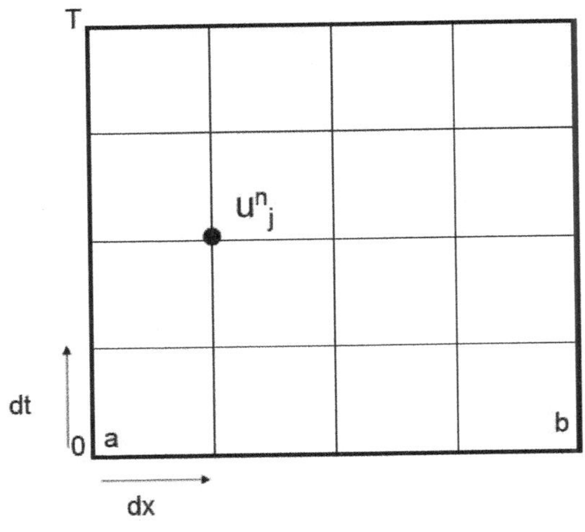

Figure 3.1. *Example of a grid*

These methods use grids for both space and time.

– Fourier space: This second class of methods works in Fourier space (the functions are expressed in terms of a Fourier basis). These methods are described as spectral methods. The idea is to decompose the functions with respect to a finite (and therefore incomplete) basis. The decomposition thus obtained is then used for the grid.

3.2. Presentation of the finite difference method

In this chapter, we will study the method of finite differences, which can be used to numerically solve first-order partial differential equations in t.

Consider the first-order hyperbolic equation in t, also know as the *transport equation*. The unknown is a function u that is defined for $(x,t) \in \mathbb{R} \times \mathbb{R}_+$. The domain of u is discretized by the grid

$$G_{h,k} = \{(x_m, t_n)/x_m = mh,\ m \in \mathbb{Z},\ t_n = nk,\ n \in \mathbb{N}\}, \qquad [3.4]$$

where h and k are strictly positive real numbers that are chosen to be as small as possible. We say that h is the spatial discretization step and k is the time step.

The function u, defined in terms of the continuous variables (x,t), takes the value $u_m^n = u(mh, nk)$ at the point $(x_m, t_n) = (mh, nk)$ of the grid $G_{h,k}$. To distinguish the continuous solution u from the result of the numerical computation, we will write v for the numerical solution, which is only defined on $G_{h,k}$: $(v_m^n)_{m\in\mathbb{Z},n\in\mathbb{Z}}$. In other words, v_m^n is the approximate solution at the point (x_m, t_n) of the PDE.

To discretize the problem, we can replace the partial derivatives by any of the following finite differences [QUA 08]:

- $\dfrac{\partial u}{\partial x}(x_m, t_n) \simeq \dfrac{u_{m+1}^n - u_m^n}{h}$, forward finite difference;
- $\dfrac{\partial u}{\partial x}(x_m, t_n) \simeq \dfrac{u_m^n - u_{m-1}^n}{h}$, backward finite difference;
- $\dfrac{\partial u}{\partial x}(x_m, t_n) \simeq \dfrac{u_{m+1}^n - u_{m-1}^n}{2h}$, centered finite difference;
- $\dfrac{\partial^2 u}{\partial x^2}(x_m, t_n) \simeq \dfrac{u_{m+1}^n - 2u_m^n + u_{m-1}^n}{h^2}$, second-order centered finite difference.

These approximations are derived from Taylor's formula. Next, we need to do the same for the partial derivatives with respect to t:

- $\dfrac{\partial u}{\partial t}(x_m, t_n) \simeq \dfrac{u_m^{n+1} - u_m^n}{k}$, forward finite difference in t.

EXAMPLE.– Consider the transport equation $u_t + cu_x = 0$ on $\mathbb{R} \times \mathbb{R}_+^*$ with the initial condition $u(x,0) = \Phi(x)$ for $x \in \mathbb{R}$, and suppose that $c > 0$.

We will use a forward method in both space and time to discretize this equation:

$$\frac{v_m^{n+1} - v_m^n}{k} + c\frac{v_{m+1}^n - v_m^n}{h} = 0. \qquad [3.5]$$

Setting $\lambda = \frac{k}{h}$ gives:

$$v_m^{n+1} = (1 + \lambda c)v_m^n - \lambda c v_{m+1}^n = (1 + \lambda c - \lambda c \mathcal{T}_h)v_m^n,$$

where $\mathcal{T}_\alpha$ is the spatial translation operator defined by $\mathcal{T}_\alpha u(x,t) = u(x - \alpha, t)$ for all $\alpha \in \mathbb{R}$, that is, $\mathcal{T}_{-h} v_m^n = v_{m+1}^n$.

By induction:

$$\begin{aligned} v_m^n &= (1 + \lambda c - \lambda c \mathcal{T}_{-h})^n v_m^0 \\ &= \sum_{p=0}^{n} C_n^p (1 + \lambda c)^{n-p} (-\lambda c \mathcal{T}_{-h})^p \Phi(x_m) \\ &= \sum_{p=0}^{n} C_n^p (1 + \lambda c)^{n-p} (-\lambda c)^p \Phi(x_m + ph). \end{aligned}$$

Thus, the domain of dependence of v at the point $(x_m, t_n) = (ih, nk)$ consists of the points x_m, $x_m + h$, $x_m + 2h, \ldots, x_m + nh$. However, we know that

$$u(x,t) = \Phi(x - ct)$$

and that the domain of dependence of $u_m^n = u(x_m, t_n)$ is simply the point $x_m - c\lambda nh$. The discrete method therefore neglects some of the properties of the solution of the PDE, and the discrete solution v_m^n is distinct from u_m^n in general: the scheme [3.5] does not converge.

Suppose that Φ was obtained experimentally, for example, in the form of a measurement $\tilde{\Phi}(x_m + ph) = \Phi(x_m + ph) + (-1)^p \epsilon$, that is, there is an error of ϵ in the value of $|\Phi|$.

The error of v_m^n would be of the order of $(1 + 2\lambda c)^n \epsilon$. For fixed λ, the error of v_m^n would grow exponentially as the number of iterations n increases. We say that the scheme [3.5] is unstable.

REMARK.– This example shows that, although it might seem straightforward to replace partial derivatives by finite differences, we need to guarantee that the solution obtained by a finite differences method converges to the solution of the partial differential equation in a sense that remains to be defined.

3.2.1. *Convergence, consistency and stability*

In the following, $\mathcal{L}$ is a linear differential operator and a first-order operator in t. We will assume that the problem

$$\begin{cases} \mathcal{L}u(x,t) = f(x,t) \text{ on } & \Omega \times \mathbb{R}_+^* \\ u(x,0) \;\; = \Phi(x) \quad \text{for } x \in \Omega \text{ open subset of } \mathbb{R} \end{cases} \qquad [3.6]$$

is well posed [RAP 05].

Replacing the partial derivatives by finite differences yields a discrete operator $\mathcal{L}_{h,k}$. This allows us to write the discretized homogeneous PDE in the form $\mathcal{L}_{h,k}v = 0$.

To account for the right-hand side of the PDE, we use a discrete operator $I_{h,k}$ that is applied to f. One example of a finite differences scheme for the PDE [3.6] is therefore:

$$\mathcal{L}_{h,k}v = I_{h,k}f. \qquad [3.7]$$

Below, we will always choose $I_{h,k}$ in such a way that $I_{h,k}f = f_m^n$, and our initial condition is simply $v_m^0 = \Phi(x_m)$ for $x_m \in \Omega$.

The following definitions give one way of classifying the various types of discrete scheme:

DEFINITION.– *A finite differences scheme is said to be explicit if we can write v_m^{n+1} as a finite linear combination of the v_i^j for $j \leq n$.*

It is said to be implicit if other values of v are required (e.g. $v_{m\pm1}^{n+1}$).

DEFINITION.– *A finite differences scheme is said to be one-step with respect to time if it only uses values of v at two points in time, for example, t_n and t_{n+1}.*

It is said to be multi-step if values of v at more than two points in time are used.

Earlier, we showed that we need a way to guarantee that the discrete solution constructed by a finite differences scheme properly represents the solution of the PDE. The next definition can help us with this.

DEFINITION.– *Let $u(x,t)$ be the solution of [3.6] and let v be a solution of the discrete scheme $\mathcal{L}_{h,k}v = f_m^n$ such that v_m^0 converges to $\Phi(x)$ whenever x_m converges to x. We say that the finite differences scheme $\mathcal{L}_{h,k}$ is convergent if v_m^n converges to $u(x,t)$ whenever (x_m, t_n) converges to (x,t) as (h,k) tends to $(0,0)$.*

REMARK.– This definition and the definitions given below only apply to one-step schemes. For multi-step schemes, we need to account for the initialization phase.

DEFINITION.– *We say that the scheme $\mathcal{L}_{h,k}v = f_m^n$ is consistent with the PDE $\mathcal{L}u = f$ if, for every function ϕ with continuity class $\mathcal{C}^\infty$, the following limit holds at every point (x_m, t_n):*

$$\lim_{(h,k)\to(0,0)} (\mathcal{L}\phi - \mathcal{L}_{h,k}\phi) = 0.$$

Consistency is a necessary condition for convergence, but is not sufficient. For example, consider again the scheme [3.5] and let ϕ be a function with continuity class $\mathcal{C}^\infty$. Then:

$$\mathcal{L}\phi = \phi_t + c\phi_x \quad \text{et} \quad \mathcal{L}_{h,k}\phi = \frac{\phi_m^{n+1} - \phi_m^n}{k} + c\frac{\phi_{m+1}^n - \phi_m^n}{k}. \qquad [3.8]$$

By Taylor's formula,

$$\phi_{m+1}^n = \phi(x_m + h, t_n) = \phi_m^n + h\phi_x(x_m, t_n) + \frac{h^2}{2}\phi_{xx}(x_m, t_n) + O(h^3),$$

$$\phi_m^{n+1} = \phi(x_m, t_n + k) = \phi_m^n + k\phi_t(x_m, t_n) + \frac{k^2}{2}\phi_{tt}(x_m, t_n) + O(k^3).$$

Therefore,

$$\mathcal{L}_{h,k}\phi(x_m, t_n) = \phi_t(x_m, t_n) + c\phi_x(x_m, t_n) + \frac{1}{2}(k\phi_{tt}(x_m, t_n)$$
$$+ch\phi_{xx}(x_m, t_n)) + O(h^2) + O(k^2),$$

and so

$$(\mathcal{L}\phi - \mathcal{L}_{h,k}\phi)(x_m, t_n) = -\frac{1}{2}(k\phi_{tt}(x_m, t_n) + ch\phi_{xx}(x_m, t_n))$$
$$+ O(h^2) + O(k^2) \xrightarrow[(h,k)\to(0,0)]{} 0.$$

This shows that the scheme [3.5] is consistent but not convergent.

DEFINITION.– *The finite differences scheme $\mathcal{L}_{h,k}v = 0$ associated with the PDE $\mathcal{L}u = 0$ is said to be stable if there exists $\Lambda \subset (\mathbb{R}_+^*)^2$ satisfying $(0,0) \in \bar{\Lambda}$ such that, for all $T > 0$, there exists a constant C_T for which the inequality*

$$h \sum_{m=-\infty}^{+\infty} |v_m^n|^2 \leq C_T h \sum_{m=-\infty}^{+\infty} |v_m^0|^2,$$

holds for every $0 \leq t \leq T$ *and* $(h,k) \in \Lambda$. *We say that* Λ *is the stability region of the scheme.*

One common example of Λ is the segment $\{(h, \lambda h) \mid 0 < h < C\}$, where C and λ are constants. Stability can be characterized by the normed space $l^2(h\mathbb{Z})$; the sequence $(v_m^n)_{m\in\mathbb{Z}}$ is an element of $l^2(h\mathbb{Z})$ if $\|v^n\|_{l^2(h\mathbb{Z})} = \left(h \sum_{m=-\infty}^{+\infty} |v_m^n|^2\right)^{1/2}$ is finite.

Then, for sufficiently small h and k in Λ, the stability criterion can be written as:

$$(\forall T > 0)(\exists C_T > 0)(\forall t_n \in [0,T]), \quad \|v^n\|_{l^2(h\mathbb{Z})} \leq C_T \|v^0\|_{l^2(h\mathbb{Z})}. \qquad [3.9]$$

Stability guarantees that, at every time $t_n \in [0,T]$, the norm of the discretized solution is bounded by the norm of the initial data multiplied by some constant factor.

This concept of numerical stability must not be confused with the separate notion of whether the well-posed problem [3.6] is stable. The latter property concerns the behavior of the solution at infinity as a function of the initial conditions, whereas the numerical stability of a scheme concerns the behavior of v on the interval $[0,T]$ as (h,k) tends to $(0,0)$ [SMA 02].

The next result explains why these notions of consistency and stability are important.

THEOREM.– *A linear scheme that is consistent with the problem [3.6] is convergent if and only if it is stable.*

The final definition of this chapter gives a way of comparing convergent schemes. Any two given convergent schemes do not necessarily achieve equivalent performance when approximating the continuous solution.

DEFINITION.– *A scheme* $L_{h,k}v = f_m^n$ *that is consistent with the problem [3.6] is said to be of order* p *in space and order* q *in time if, for every function* ϕ *with continuity class* C^∞*:*

$$E_{h,k}\phi = L_{h,k}\phi - L\phi = O(h^p) + O(k^q). \qquad [3.10]$$

If so, we say that the scheme is of order (p,q)*, and* $E_{h,k}$ *is called the truncation error.*

3.2.2. *Courant–Friedrichs–Lewy condition*

Below, we will consider several examples of discrete schemes for the transport problem with speed $c \in \mathbb{R}^*$:

$$\begin{cases} u_t + cu_x = 0 & \text{on } \mathbb{R} \times \mathbb{R}_+^* \\ u(x,0) = \Phi(x) & \text{for } x \in \mathbb{R} \end{cases} \qquad [3.11]$$

– The forward scheme in time and space [3.5] is of the form $v_m^{n+1} = \alpha v_m^n + \beta v_{m+1}^n$. This scheme satisfies

$$\|v^{n+1}\|_{l^2(h\mathbb{Z})}^2 = h \sum_{m\in\mathbb{Z}} |\alpha v_m^n + \beta v_{m+1}^n|^2 \leq (|\alpha| + |\beta|)^2 \|v^n\|_{l^2(h\mathbb{Z})}^2.$$

Hence, it can only be stable if $|\alpha| + |\beta| \leq 1$.

In particular, for the scheme [3.5], $|1 + c\alpha| + |c\alpha| \leq 1$, so it is stable for $-1 \leq c\alpha \leq 0$.

– The Lax–Friedrichs scheme is centered in space and forward in time, but uses a centered average to approximate v_m^n:

$$\frac{v_m^{n+1} - \frac{1}{2}(v_{m+1}^n + v_{m-1}^n)}{k} + c\frac{v_{m+1}^n - v_{m-1}^n}{2h} = 0. \qquad [3.12]$$

This scheme is consistent, and is of the form $v_m^{n+1} = \alpha v_{m+1}^n + \beta v_{m-1}^n$.

It can be shown that these schemes are in fact stable whenever $|\alpha| + |\beta| \leq 1$, and in particular the Lax–Friedrichs scheme [3.12] is stable whenever $|c\lambda| \leq 1$ [TRE 96].

The results stated above can be generalized as follows:

THEOREM.– *Suppose that $v_m^{n+1} = \alpha v_{m-1}^n + \beta v_m^n + \gamma v_{m+1}^n$ is an explicit scheme for the PDE [3.11]. Then, if the ratio $\frac{k}{h} = \lambda$ is constant, the scheme can only be stable if the Courant–Friedrichs–Lewy (CFL) condition holds: $|c\lambda| \leq 1$.*

THEOREM.– *Explicit schemes for the PDE [3.11] cannot be both consistent and unconditionally stable.*

EXAMPLE.– The above theorem does not extend to implicit schemes. For example, consider the following scheme, which is forward in time and backward in space, at time t_{n+1}:

$$\frac{v_m^{n+1} - v_m^n}{k} + c\frac{v_m^{n+1} - v_{m-1}^{n+1}}{h} = 0.$$

By writing this scheme in the form $(1 + c\lambda)v_m^{n+1} = v_m^n + c\lambda v_{m-1}^{n+1}$, we can show that, for $c > 0$ and for all $\lambda \in \mathbb{R}_+$:

$$\|v^{n+1}\|^2_{l^2(h\mathbb{Z})} \leq \|v^n\|^2_{l^2(h\mathbb{Z})}.$$

Therefore, this scheme is unconditionally stable for $c > 0$.

3.2.3. *Von Neumann stability analysis*

Von Neumann stability analysis uses Fourier analysis to study the behavior of discrete schemes.

Consider a linear one-step discrete scheme $L_{h,k}v = 0$ with constant coefficients. Taking the Fourier transform and rearranging terms leads to the relation:

$$\hat{v}^{n+1}(\xi) = g(\xi, h, k)\hat{v}^n(\xi). \qquad [3.13]$$

The factor $g(\xi, h, k)$ is said to be the amplification factor: the modulus $|g|$ of g is the amplification, and the phase $\arg g$ of g is the phase offset introduced at each frequency of the discrete solution when increasing by one time step k. Iterating this gives the relation:

$$\hat{v}^n(\xi) = g(\xi, h, k)^n \hat{v}^0(\xi).$$

EXAMPLE.– Consider the transport equation $u_t + cu_x = 0$ for $c > 0$, discretized using [3.5], a forward scheme in both time and space.

Then $g(\xi, h, k) = 1 + c\lambda - c\lambda e^{ih\xi}$. For fixed λ, this only depends on $h\xi$, and

$$|g(h\xi)|^2 = g(h\xi)g(\bar{h}\xi) = 1 + 4c\lambda \sin^2(h\xi/2).$$

However,

$$|\hat{v}^n(\xi)| = |g(h\xi)|^n |\hat{v}^0(\xi)|,$$

and $|g(h\xi)| = 1$ if and only if $h\xi = 0$; $|g(h\xi)| > 1$. We can use this fact to deduce that the scheme is unstable.

Let $(h_0, k_0) \in (\mathbb{R}_+^*)^2$, and $K \in \mathbb{R}_+^*$. There exists $(h, k) \in]0, h_0] \times]0, k_0]$ and $(\xi_1, \xi_2) \in]0, \pi/h[^2$ such that

$$|g(h\xi)| \geq 1 + Kh \text{ for all } \xi \in [\xi_1, \xi_2].$$

We define $\hat{v}^0$ to satisfy

$$\hat{v}^0(\xi) = \begin{cases} (\xi_2 - \xi_1)^{-1/2} & \text{if } \xi \in [\xi_1, \xi_2] \\ 0 & \text{if } \xi \notin [\xi_1, \xi_2] \end{cases}$$

and $\|v^0\|_{l^2(h\mathbb{Z})} = 1$. Thus,

$$\begin{aligned} \|v^n\|^2_{l^2(h\mathbb{Z})} &= \|\hat{v}^n\|^2_{L^2([-\pi/h,\pi/h])} \\ &= \int_{-\frac{\pi}{h}}^{+\frac{\pi}{h}} |g(h\xi)|^{2n} |\hat{v}^0(\xi)|^2 d\xi \\ &= \int_{\xi_1}^{\xi_2} |g(h\xi)|^{2n} \frac{1}{\xi_2 - \xi_1} d\xi \\ &\geq (1 + Kk)^{2n} \geq \frac{1}{2} e^{2KT} \|v^0\|^2_{l^2(h\mathbb{Z})} \end{aligned}$$

for $n \simeq T/k$ and k_0 sufficiently small.

Since K may be chosen to be arbitrarily large, the stability criterion (S) is not satisfied and the scheme [3.5] is unstable and hence divergent for $c > 0$.

In general, the following result holds:

THEOREM.– *A linear one-step scheme $L_{h,k}v = 0$ with constant coefficients is stable if and only if there exists a constant $K \in \mathbb{R}_+$ together with a set $\Lambda \subset (\mathbb{R}^*_+)^2$ with $(0, 0) \in \bar{\Lambda}$ such that*

$$|g(\xi, h, k)| \leq 1 + Kk$$

for all $(h, k) \in \Lambda$. If the amplification factor only depends on $h\xi$, the scheme is stable if and only the following condition holds:

$$|g(h\xi))| \leq 1.$$

3.3. Hyperbolic equations

Hyperbolic problems behave differently from elliptic or parabolic equations. For example, they can exhibit a special phenomenon known as shocks. We will study three examples of hyperbolic equations.

Two of these examples are linear hyperbolic equations:

– the transport equation;

– the wave equation.

These two equations can be written in the form:

$$\frac{\partial^2 u}{\partial t^2} - \nabla(a\nabla u) + cu = f.$$

Finally, as a prototype for nonlinear hyperbolic equations, we will also consider Burgers' equation:

$$\frac{\partial u}{\partial t} + \frac{\partial f(u)}{\partial x} = 0.$$

3.3.1. *Key results*

Shocks or *shock waves* are singularities in the solution of a PDE. Some linear hyperbolic equations admit shocks, in which case these shocks are necessarily specified in the initial conditions or the boundary conditions, and propagate along the characteristics of the differential equation.

For nonlinear hyperbolic equations, shocks that are not present in the data (either the initial conditions or the boundary conditions) can nonetheless arise if the characteristics intersect.

EXAMPLE.– Consider the transport equation

$$\frac{\partial u}{\partial t} + c\frac{\partial u}{\partial x} = \rho$$

with boundary conditions:

$$\begin{cases} u(x,0) = 0 \text{ if } x < \rho \\ u(x,0) = \rho \text{ if } x > \rho \end{cases}$$

The function $u(x,t)$ is discontinuous at $(\rho, 0)$, assuming that $\rho \neq 0$. The characteristics are therefore defined by the equation

$$\frac{dt}{1} = \frac{dx}{c} = \frac{du}{\rho}.$$

Each of these characteristics is a straight line. The equation of the characteristic that passes through the point $(x_0, 0)$ is

$$x = ct + x_0.$$

The solutions of the PDE are given by:

$$\begin{cases} u(x,0) = \frac{\rho(x-x_0)}{c} & \text{if } x_0 < \rho \\ u(x,0) = \frac{\rho(x-x_0)}{c} + \rho & \text{if } x_0 > \rho \end{cases}$$

Along any given characteristic,

$$\lim_{x_0 \to \rho^-} u(x,t) \neq \lim_{x_0 \to \rho^+} u(x,t).$$

Hence, the function $u(x,t)$ is discontinuous along the characteristics of the PDE.

EXAMPLE.– Consider the wave equation

$$\frac{\partial^2 u}{\partial t^2} = c^2 \frac{\partial^2 u}{\partial x^2}$$

with boundary conditions:

$$\begin{cases} u(x,t) = 0 & \text{if } x = ct \text{ and } x \leq 1 \\ u(x,t) = x - 1 & \text{if } x = ct \text{ and } x \geq 1 \\ u(x,t) = x & \text{if } x = -ct \end{cases}$$

The solution of this equation is:

$$\begin{cases} u(x,t) = \frac{1}{2}(x - ct) & \text{if } x \leq 1 \\ u(x,t) = x - 1 & \text{if } x \geq 1 \end{cases}$$

This solution is said to be a *weak solution* because the function $u(x,t)$ is continuous, but is not differentiable at the point $(x = 1, t = \frac{1}{c})$. The boundary condition is not differentiable at this point: the singularity is inherited by the solution.

EXAMPLE.– Consider the following nonlinear hyperbolic equation. It was studied by J.M. Burgers:

$$\frac{\partial u}{\partial t} + u\frac{\partial u}{\partial x} = 0.$$

The Cauchy problem

$$\begin{cases} \frac{\partial u}{\partial t} + u\frac{\partial u}{\partial x} = 0 \text{ if } x \in \mathbb{R}, \text{ and } t > 0 \\ u(x,0) = u_0(x) \end{cases}$$

has a solution $u(x,t)$ that is parametrically described by the characteristic equation (D_λ):

$$x = u_0(\lambda)t + \lambda.$$

The solution is constant along each of these curves $u = u_0(\lambda)$.

The next series of examples explores the question of whether a problem is well posed or ill posed, and studies the conditions under which shocks can arise.

Suppose that the function u_0 is the Heaviside function:

$$\begin{cases} u_0(\lambda) = 0 \text{ if } \lambda \leq 0 \\ u_0(\lambda) = 1 \text{ if } \lambda > 0 \end{cases}$$

With these assumptions, one possible solution is given by:

$$\begin{cases} u(x,t) = 0 \ \text{ if } x \leq 0 \\ u(x,t) = 1 \ \text{ if } 0 < t \leq x \\ u(x,t) = \frac{x}{t} \text{ if } 0 \leq x \leq t \end{cases}$$

This solution is continuous even though the initial conditions are discontinuous. However, the system has another solution, which features a shock:

$$\begin{cases} u(x,t) = 0 \text{ if } x \leq t/2 \\ u(x,t) = 1 \text{ if } x > t/2 \end{cases}$$

The shock propagates along the straight line described by the equation $x = \frac{t}{2}$.

We can impose an additional condition, the so-called entropy condition, to eliminate this solution:

$$\forall x,\ \forall t > 0, \quad u(x-0,t) \geq u(x+0,t).$$

This condition accepts shocks that decrease u and rejects shocks that increase u. Consider the initial values:

$$\begin{cases} u_0(\lambda) = 0 \qquad \text{if } \lambda \leq 0 \\ \\ u_0(\lambda) = -\lambda^2 \text{ if } \lambda > 0 \end{cases}$$

With these values, the characteristics intersect. Writing

$$f(x,t,\lambda) = x - u_0(\lambda) - \lambda,$$

the envelope of the characteristics is described by the parametric system $f = 0$ and $\partial f/\partial\lambda = 0$, or in other words the system:

$$\begin{cases} t = \dfrac{-1}{u_0(\lambda)} \\ x = \lambda - \dfrac{u_0(\lambda)}{u_0'(\lambda)} \end{cases}$$

Here, this is the branch of a hyperbola described by the equation $t = \frac{1}{4}x$.

We know that the solution is constant and equal to the slope $u_0(\lambda)$ on each characteristic. Therefore, if the characteristics intersect, the function $u(x,t)$ takes two or more values. This is inadmissible, so the problem is ill posed. Furthermore, if the derivatives of u are discontinuous, the PDE is no longer defined, since the derivatives do not exist.

DEFINITION.– *We say that u has a discontinuity of first category on the curve C if u is not continuous, but u and its derivatives have left and right derivatives that are continuous functions in curvilinear coordinates on C.*

If u is a piecewise C^1 function with discontinuities of the first category on the curve C, then it can be shown that the slope of the shock is equal to the average of the values on either side of the shock:

$$\left.\frac{dx}{dt}\right|_C = \frac{u^- + u^+}{2}. \qquad [3.14]$$

For the generalized Burgers equation, the Rankine–Hugoniot conditions are used to describe the requirement that the flow $f(x)$ should be continuous at the discontinuities of the curve $x = x(t)$.

If u_0 is a bounded measurable function, it can be shown that the Cauchy problem, reformulated as:

$$\begin{cases} \dfrac{\partial u}{\partial t} + u\dfrac{\partial u}{\partial x} = 0 & \text{if } x \in \mathbb{R}, \text{ and } t > 0 \\ u(x,0) = u_0(x) & \\ \dfrac{dx}{dt}|_C = \dfrac{u^- + u^+}{2} & \text{Rankine-Hugoniot condition} \\ \forall x,\ \forall t > 0,\ u(x-0,t) \geq u(x+0,t) & \text{entropy condition} \end{cases}$$

has a unique solution.

This problem can equivalently be stated in a variational form:

$$\begin{cases} \forall \phi \in \mathcal{C}^1(\mathbb{R}^2) \text{ of bounded support} \\ \int_{-\infty}^{+\infty} dx \int_0^{+\infty} (u\frac{\partial \phi}{\partial t} + \frac{u^2}{2}\frac{\partial \phi}{\partial x})dt + \int_{-\infty}^{+\infty} u_0(x)\phi(x,0)dx = 0 \\ \forall x,\ \forall t > 0, \quad u(x-0,t) \geq u(x+0,t) \end{cases}$$

3.3.2. *Numerical schemes for solving the transport equation*

This section presents a few of the most commonly used schemes for solving the homogeneous problem [3.11] when $c \in \mathbb{R}^*$.

3.3.2.1. *Spatially centered scheme*

The spatially centered scheme is defined as follows:

$$\frac{v_m^{n+1} - v_m^n}{k} + c\frac{v_{m+1}^n - v_{m-1}^n}{2h} = 0.$$

This scheme is consistent and has order (2,1). Writing

$$v_m^{n+1} = v_m^n - \frac{c\lambda}{2}(v_{m+1}^n - v_{m-1}^n),$$

where $\lambda = \frac{k}{h}$, it can be shown that the amplification factor is $g(h\xi) = 1 - ic\lambda \sin(h\xi)$ and $|g|^2 = 1 + c^2\lambda^2 \sin^2(h\xi) \geq 1$. Hence, the scheme is unstable.

3.3.2.2. *Upwind scheme*

If we test the sign of c, we obtain a scheme that uses either v_{m-1}^n or v_{m+1}^n depending on the direction of advection (hence the name "upwind"). This scheme can be stated as follows:

$$\frac{v_m^{n+1} - v_m^n}{k} + \left(\frac{c - |c|}{2}\right)\frac{v_{m+1}^n - v_m^n}{h} + \left(\frac{c + |c|}{2}\right)\frac{v_m^n - v_{m-1}^n}{h} = 0.$$

Equivalently:

$$\frac{v_m^{n+1} - v_m^n}{k} + c\frac{v_{m+1}^n - v_{m-1}^n}{2h} - |c|\frac{h}{2}\frac{v_{m+1}^n - 2v_m^n + v_{m-1}^n}{h^2} = 0.$$

Hence, this is an explicit centered scheme with a correction term originating from the discretization of $(|c|h/2)u_{xx}$.

This scheme is consistent, of order (1,1), and $g(h\xi) = 1 - 2|c|\lambda \sin^2(h\xi/2) - ic\lambda \sin(h\xi)$. The relation $|g(h\xi)|^2 = 1 - 4|c|\lambda(1 - |c|\lambda) \sin^2(h\xi) \leq 1$ holds if and only if $|c|\lambda \leq 1$. Note that the numerical dissipation or viscosity term $(|c|h/2)u_{xx}$ stabilizes the spatially centered scheme, but reduces its order.

3.3.2.3. *Lax–Friedrichs scheme*

The Lax–Friedrichs scheme can be stated as follows:

$$\frac{v_m^{n+1} - \frac{1}{2}(v_{m+1}^n + v_{m-1}^n)}{k} + c\frac{v_{m+1}^n - v_{m-1}^n}{2h} = 0.$$

If ϕ is regular, it is possible to show that

$$\mathcal{L}_{h,k}\phi - \mathcal{L}\phi = \frac{k}{2}\phi_{tt} - \frac{h^2}{2k}\phi_{xx} + O(k^2 + h^4/k) + \frac{ch^2}{6}\phi_{xxx} + O(h^4).$$

This scheme is consistent, provided that $\frac{h^2}{k} \to 0$ as $h, k \to 0$. If $k = \lambda h$, where $\lambda \in \mathbb{R}_+^*$ is fixed, then the Lax–Friedrichs scheme is consistent of order one.

The amplification factor is $g(h\xi) = \cos(h\xi) + ic\lambda \sin(h\xi)$, and $|g|^2 = \cos^2(h\xi) + c^2\lambda^2 \sin^2(h\xi) \leq 1$ if and only if $|c|\lambda \leq 1$.

The scheme can be rewritten in the form:

$$\frac{v_m^{n+1} - v_m^n}{k} + c\frac{v_{m+1}^n - v_{m-1}^n}{2h} - \frac{1}{2}\frac{v_{m+1}^n - 2v_m^n + v_{m-1}^n}{k} = 0.$$

Note that the Lax–Friedrichs scheme is equivalent to discretizing the following PDE by finite centered differences:

$$u_t + cu_x - \frac{h^2}{2k}u_{xx} = 0.$$

The phenomenon of *numerical viscosity* is again present, with a stabilizing effect on the scheme.

3.3.2.4. *Lax–Wendroff scheme*

The Lax–Wendroff scheme uses Taylor expansions to discretize the PDE. If u is a regular solution of the PDE, then:

$$u_m^{n+1} = u_m^n + k\frac{\partial u}{\partial t} + \frac{k^2}{2}\frac{\partial^2 u}{\partial t^2} + O(k^2).$$

However,

$$u_t = -cu_x \quad \text{and} \quad u_{tt} = (-cu_x)_t = -cu_{xt} = c^2 u_{xx},$$

so

$$u_m^{n+1} = u_m^n - ck\frac{\partial u}{\partial x} + c^2\frac{k^2}{2}\frac{\partial^2 u}{\partial x^2} + O(k^2).$$

By using spatially centered discretization to approximate u_x and u_{xx}, we obtain the following discrete scheme:

$$v_m^{n+1} = v_m^n - ck\frac{v_{m+1}^n - v_{m-1}^n}{2h} + c^2\frac{k^2}{2}\frac{v_{m+1}^n - 2v_m^n + v_{m-1}^n}{h^2}.$$

This scheme is consistent and of order (2,1), since

$$L_{h,k}\phi - L\phi = \frac{k}{2}\phi_{tt} - \frac{c^2 k}{2}\phi_{xx} + \frac{c^2 kh^2}{4!}\phi_{xxxx} + O(h^3 + k^2).$$

Setting $\lambda = \frac{k}{h}$, we have that:

$$v_m^{n+1} = v_m^n - \frac{c\lambda}{2}(v_{m+1}^n - v_{m-1}^n) + \frac{c^2\lambda^2}{2}(v_{m+1}^n - 2v_m^n + v_{m-1}^n).$$

It can be shown that the amplification factor is

$$g(h\xi) = 1 - 2c^2\lambda^2\sin^2(h\xi/2) - ic\lambda\sin(h\xi)$$

and

$$|g|^2 = 1 - 4c^2\lambda^2(1 - c^2\lambda^2)\sin^4(h\xi/2).$$

This scheme is therefore stable whenever the CFL condition $|c\lambda| \leq 1$ holds.

3.3.2.5. *Leapfrog scheme*

The so-called “leapfrog” scheme is centered in both space and time:

$$\frac{v_m^{n+1} - v_m^{n-1}}{2k} + c\frac{v_{m+1}^n - v_{m-1}^n}{2h} = 0,$$

or, alternatively,

$$v_m^{n+1} = v_m^{n-1} - c\lambda(v_{m+1}^n - v_{m-1}^n).$$

This is a two-step scheme, so we need initial conditions for both v_m^0 and v_m^1. In practice, a one-step scheme is used to initialize multi-step schemes. Since

$$\mathcal{L}_{h,k}\phi - \mathcal{L}\phi = \frac{k^2}{6}\phi_{ttt} - \frac{ch^2}{6}\phi_{xxx} + O(h^2 + k^2),$$

the leapfrog scheme is consistent and of order $(2,2)$. We cannot apply von Neumann analysis to this scheme directly, but it nonetheless has the following properties:

– if $c\lambda < 1$, the scheme is stable;

– if $c\lambda \geq 1$, the scheme is not stable.

3.3.3. *Wave equation*

The wave equation is the canonical example of a second-order linear hyperbolic equation. Its statement is:

$$\frac{\partial^2 u}{\partial t^2} = c^2 \frac{\partial^2 u}{\partial x^2}. \qquad [3.15]$$

Any function of the form

$$u(x,t) = f(x + ct) + g(x - ct)$$

is a solution, where f and g are arbitrary C^2 functions, representing the sum of a forward-propagating wave and a backward-propagating wave.

Consider the following Cauchy problem:

$$\begin{cases} \frac{\partial^2 u}{\partial t^2} & = c^2 \frac{\partial^2 u}{\partial x^2} \text{ if } x \in \mathbb{R} \text{ and } t > 0, \\ u(x,0) & = u_0(x) \\ \frac{\partial u}{\partial t}(x,0) & = u_1(x) \end{cases} \qquad [3.16]$$

If u_0 has continuity class C^p and u_1 has continuity class C^{p-1}, then this Cauchy problem has a classical solution with continuity class C^p:

$$u(x,t) = u_0(x + ct)/2 + u_0(x - ct)/2 + \frac{1}{2c}\int_{x-ct}^{x+ct} u_1(\tau)d\tau. \qquad [3.17]$$

3.3.3.1. *Theta-scheme method*

Consider the wave equation,

$$\frac{\partial^2 u}{\partial t^2} - c^2 \frac{\partial^2 u}{\partial x^2} = 0.$$

The θ-scheme associated with the wave equation can be stated as

$$\frac{u_{i,j+1} - 2u_{i,j} + u_{i,j-1}}{(\Delta t)^2} - c^2 \frac{\theta A u_{i,j+1} + (1-2\theta) A u_{i,j} + \theta A u_{i,j-1}}{(\Delta x)^2} = 0,$$

where

$$A u_{i,j} = u_{i+1,j} - 2u_{i,j} + u_{i-1,j}.$$

This scheme is explicit if $\theta = 0$ and implicit otherwise. For $0 \leq \theta \leq \frac{1}{4}$, the scheme is stable whenever

$$c\frac{\Delta t}{\Delta x} \leq \frac{1}{\sqrt{1-4\theta}}.$$

For $\frac{1}{4} \leq \theta \leq 1$, the scheme is universally stable.

3.3.3.2. *Lax scheme*

The wave equation can be written in the form of a system:

$$\begin{cases} \dfrac{\partial v}{\partial t} = c\dfrac{\partial w}{\partial x} \\ \dfrac{\partial w}{\partial t} = \dfrac{\partial v}{\partial x} \end{cases}$$

The Lax scheme can then be stated as:

$$\begin{cases} v_{i,j+1} = \frac{1}{2}(v_{i+1,j} + v_{i-1,j}) + c\frac{\Delta t}{2\Delta x}(w_{i+1,j} - v_{i-1,j}) \\ w_{i,j+1} = \frac{1}{2}(w_{i+1,j} + w_{i-1,j}) + c\frac{\Delta t}{2\Delta x}(v_{i+1,j} - v_{i-1,j}) \end{cases}$$

This scheme is a first-order, two-level scheme that is stable whenever the CFL condition holds:

$$c\frac{\Delta t}{\Delta x} \leq 1.$$

3.3.3.3. *Leapfrog scheme*

The leapfrog scheme can also be applied to the wave equation expressed as a first-order system:

$$\begin{cases} v_{i,j+1} &= v_{i,j} + c\frac{\Delta t}{\Delta x}(w_{i+1,j} - w_{i-1,j}) \\ \\ w_{i,j+1} &= w_{i,j-1} + c\frac{\Delta t}{\Delta x}(v_{i+1,j} - v_{i-1,j}) \end{cases}$$

This is an explicit second-order, three-level scheme that is stable whenever the CFL condition is satisfied.

3.3.3.4. *Lax–Wendroff scheme*

The Lax–Wendroff scheme with $\lambda = \Delta t/\Delta x$ can be stated as follows:

$$\begin{cases} v_{i,j+1} &= v_{i,j} + c\frac{\lambda}{2}(w_{i+1,j} - w_{i-1,j}) + \frac{c^2\lambda^2}{2}(v_{i+1,j} - 2v_{i,j} + v_{i-1,j}) \\ \\ w_{i,j+1} &= w_{i,j} + c\frac{\lambda}{2}(v_{i+1,j} - v_{i-1,j}) + \frac{c^2\lambda^2}{2}(w_{i+1,j} - 2w_{i,j} + w_{i-1,j}) \end{cases}$$

This scheme is stable whenever the CFL condition holds.

3.3.4. **Burgers equation**

The equation studied by J.M. Burgers,

$$\frac{\partial u}{\partial t} + u\frac{\partial u}{\partial x} = 0, \qquad [3.18]$$

can be generalized to

$$\frac{\partial u}{\partial t} + \frac{\partial f(u)}{\partial x} = 0. \qquad [3.19]$$

The Cauchy problem associated with this equation is:

$$\begin{cases} \dfrac{\partial u}{\partial t} + u\dfrac{\partial u}{\partial x} &= 0 \qquad \text{if } x \in \mathbb{R} \text{ and } t > 0 \\ u(x,0) &= u_0(x) \end{cases} \qquad [3.20]$$

This problem admits a solution $u(x,t)$ that is parametrically described by the characteristic equation (D_λ):

$$x = u_0(\lambda)t + \lambda.$$

Whenever f is a convex function, the Burgers equation with initial conditions

$$u(x,0) = \begin{cases} u_g \text{ if } x < 0 \\ u_d \text{ if } x > 0 \end{cases}$$

has a unique solution $u(x,t) = w(\frac{x}{t}, u_g, u_d)$, where w is the function known as the *Riemann solver* and g is the function that satisfies $f'(g(x)) = x$. The Riemann solver is defined by:

$$w(y,u,v) = \begin{cases} u & \text{if } y < f'(u) \\ g(y) & \text{if } f'(u) < y < f'(v) \\ v & \text{if } y > f'(v) \end{cases}$$

Below, we will assume that f is convex and has continuity class $\mathcal{C}^2$.

3.3.4.1. *Lax–Friedrichs scheme*

The Lax–Friedrichs scheme can be stated as:

$$\frac{u_{i,j+1} - \frac{1}{2}(u_{i-1,j} + u_{i+1,j})}{\Delta t} + \frac{f(u_{i+1,j}) - f(u_{i-1,j})}{2\Delta x} = 0. \qquad [3.21]$$

This is an explicit first-order, two-level scheme in time that is stable if the CFL condition is satisfied:

$$\frac{\Delta t}{\Delta x} \sup |f'(u_{i,j})| \leq 1.$$

3.3.4.2. *Leapfrog scheme*

The leapfrog scheme can be applied to the Burgers equation as follows:

$$u_{i,j+1} = u_{i,j-1} - \frac{\Delta t}{2\Delta x}(f(u_{i+1,j}) - f(u_{i-1,j})). \qquad [3.22]$$

This is an explicit second-order, three-level scheme that is stable when the CFL condition holds.

3.3.4.3. *Lax–Wendroff scheme*

Consider the Taylor expansion up to second order of the Burgers equation:

$$u(x, t + \Delta t) = u(x, t) + \Delta\frac{\partial u}{\partial t}(x, t) + \frac{\Delta t^2}{2}\frac{\partial^2 u}{\partial t^2}(x, t) + O(\Delta t^3).$$

After discretization, this becomes:

$$u_{i,j+1} = u_{i,j} + \Delta t\left(\frac{\partial u}{\partial t}\right)_{i,j} + \frac{\Delta t^2}{2}\left(\frac{\partial^2 u}{\partial t^2}\right)_{i,j} + O(\Delta t^3).$$

Now discretize by centered differences:

$$\left(\frac{\partial u}{\partial t}\right)_{i,j} = -\left(\frac{\partial f(u)}{\partial x}\right)_{i,j} = \frac{-f(u_{i+1,j}) + f(u_{i-1,j})}{2\Delta x} + O(\Delta x^2).$$

Note that

$$\frac{\partial^2 u}{\partial t^2} = \frac{\partial}{\partial t}\left(-\frac{\partial f(u)}{\partial x}\right) = -\frac{\partial}{\partial x}\left(f'(u)\frac{\partial u}{\partial t}\right) = \frac{\partial}{\partial x}\left(f'(u)\frac{\partial f(u)}{\partial x}\right).$$

Given $\theta \in [0, 1]$, if we set $\partial_x f(u) = \partial f(u)/\partial x$, then:

$$\begin{aligned}\frac{\partial}{\partial x}(f'(u)\partial_x f(u)) = f'(u(x + \theta\Delta x, t))&\frac{f(u(x + \Delta x, t)) - f(u(x - \Delta x, t))}{\Delta x^2}\\ - f'(u(x + (\theta - 1)\Delta x, t))&\frac{f(u(x, t)) - f(u(x - \Delta x, t))}{\Delta x^2}\\ + O(\Delta x).&\end{aligned}$$

In the above expression, we replaced

$$\frac{\partial f(u)}{\partial x} = \frac{f(u(x + \Delta x, t)) - f(u(x - \Delta x, t))}{\Delta x^2},$$

which gives the discretized expression

$$\begin{aligned}\frac{\partial}{\partial x}\left(f'(u)\frac{f(u)}{\partial x}\right)_{i,j} &= \frac{f'(u_{i+\theta,j})}{\Delta x^2}(f(u_{i+1,j}) - f(u_{i,j}))\\ &- \frac{f'(u_{i+(\theta-1),j})}{\Delta x^2}(f(u_{i,j}) - f(u_{i-1,j})).\end{aligned}$$

Finally, by choosing $\theta = \frac{1}{2}$, we obtain the Lax–Wendroff scheme:

$$u_{i,j+1} = u_{i,j} - \frac{\lambda^2}{2}(f(u_{i+1,j}) - f(u_{i-1,j})) + \frac{\lambda^2}{2} f'(u_{u+1/2,j})(f(u_{i+1,j}) - f(u_{i,j})) - \frac{\lambda^2}{2} f'(u_{i-1/2,j})(f(u_{i,j}) - f(u_{i-1,j})),$$

where $\lambda = \frac{\Delta t}{\Delta x}$ and

$$f'(u_{i\pm 1/2,j}) = \frac{f'(u_{i,j}) + f'(u_{i\pm 1,j})}{2}.$$

The Lax–Wendroff scheme is an explicit second-order, two-level scheme that is stable when the CFL condition holds.

3.3.4.4. *Engquist–Osher scheme*

This scheme is used for the generalized Burgers equation:

$$\frac{\partial u}{\partial t} + \frac{\partial f(u)}{\partial x} = 0.$$

The Engquist–Osher scheme is a generalization of the Lax–Wendroff scheme:

$$u_{i,j+1} = u_{i,j} - \lambda(\Phi(u_{i,j}, u_{i+1,j}) - \Phi(u_{i-1,j}, u_{i,j})), \qquad [3.23]$$

where $\lambda = \frac{\Delta t}{\Delta x}$, and the *numerical flux* $\Phi(u, v)$ is defined by

$$\Phi(u, v) = \frac{1}{2}(f(v) - f(u)) - \frac{\lambda}{2} \int_u^v |f'(\tau)| d\tau.$$

The integral term discretizes the numerical viscosity. This scheme is of first order and is stable when the CFL condition is satisfied.

3.3.4.5. *Godunov scheme*

The Godunov scheme is also used for the generalized Burgers equation:

$$\frac{\partial u}{\partial t} + \frac{\partial f(u)}{\partial x} = 0. \qquad [3.24]$$

Here, we introduce the *numerical flux* $\Phi(u, v)$ in terms of the Riemann solver $w(0, u, v)$:

$$\Phi(u, v) = w(0, u, v) = \begin{cases} f(u) & \text{if } f'(u) > 0 \\ f \circ g(0) & \text{if } f'(u) < 0 \text{ and } f'(v) > 0 \\ f(v) & \text{if } f'(v) < 0 \end{cases}$$

where g is the function that satisfies $f'(g(x)) = x$. The Godunov scheme can be stated as:

$$u_{i,j+1} = u_{i,j} - \frac{\Delta t}{\Delta x}(\Phi(u_{i,j}, u_{i+1,j}) - \Phi(u_{i-1,j}, u_{i,j})). \quad [3.25]$$

This is a first-order scheme that is stable when the CFL condition holds.

3.3.4.6. *Lerat–Peyret scheme*

Again for the generalized Burgers equation:

$$\frac{\partial u}{\partial t} + \frac{\partial f(u)}{\partial x} = 0.$$

The Lerat–Peyret schemes S_α^β are a family of second-order schemes indexed by the parameters α and β:

– Choosing $\alpha = 1$ and $\beta = 0$ yields the Mac–Cormak scheme.

– Choosing $\alpha = \beta = 1/2$ yields the Richtmeyer scheme.

These schemes follow a prediction–correction approach to solving the Burgers equation:

– the predictor is

$$p_i = (1 - \beta)u_{i,j} + \beta u_{i+1,j} - \alpha\frac{\Delta t}{\Delta x}(f(u_{i+1,j}) - f(u_{i,j})). \quad [3.26]$$

– the corrector is

$$\begin{aligned} u_{i,j+1} = u_{i,j} - \frac{\Delta t}{2\alpha\Delta x}((\alpha - \beta)f(u_{i+1,j}) + (2\beta - 1)f(u_{i,j}) \\ + (1 - \alpha - \beta)f(u_{i-1,j}) + f(p_i) - f(p_{i-1})). \quad [3.27] \end{aligned}$$

Lerat–Peyret schemes are stable when the CFL condition holds.

3.4. Elliptic equations

3.4.1. *Poisson equation*

3.4.1.1. *Richardson–Liebmann method*

This method discretizes the Poisson equation

$$\frac{\partial^2 u}{\partial x^2} + \frac{\partial^2 u}{\partial y^2} = f \quad [3.28]$$

and its boundary conditions by the expression

$$\frac{u_{i+1,j} - 2u_{i,j} + u_{i-1,j}}{(\Delta x)^2} + \frac{u_{i,j+1} - 2u_{i,j} + u_{i,j-1}}{(\Delta y)^2} = f_{i,j}. \qquad [3.29]$$

When the discretization step is equal in both x and y, that is, $h = \Delta x = \Delta y$, the above scheme can be rewritten more simply as

$$u_{i+1,j} + u_{i-1,j} + u_{i,j+1} + u_{i,j-1} - 4u_{i,j} = h^2 f_{i,j}. \qquad [3.30]$$

This gives a system of equations whose unknowns are the values $u_{i,j}$ of the function u at each of the nodes of the discretization mesh. A matrix method is used to solve this system:

– This method is called the Liebmann method when the Gauss–Seidel method is used to solve the system.

– It is known as the Richardson method when the system is solved using the Jacobi method.

3.4.1.2. *Relaxation methods*

Relaxation methods rewrite the usual discretized form of the Poisson equation as a linear system, then solve it with a relaxation method. At the k-th iteration, the method computes

$$u_{i,j}^{(k+1)} = (1-\omega)u_{i,j}^{(k)} + \omega\xi_{i,j}^{(k)},$$

where

$$\xi_{i,j}^{(k)} = \frac{1}{4}(f_{i,j}h^2 - u_{i+1,j}^{(k)} - u_{i-1,k}^{(k)} - u_{i,j-1}^{(k)}).$$

3.4.1.3. *Fast Fourier Transform method*

This method considers the Poisson equation in the discretized form

$$u_{i+1,j} + u_{i-1,j} + u_{i,j+1} + u_{i,j-1} - 4u_{i,j} = h^2 f_{i,j}$$

and then applies a Fourier transform to obtain

$$(e^{2i\pi m/I} + e^{-2i\pi m/I} + e^{2i\pi n/J} + e^{-2i\pi n/J} - 4)\hat{u}_{m,n} = h^2 \hat{f}_{m,n}, \qquad [3.31]$$

or, equivalently,

$$2(\cos 2\pi m/I + \cos 2\pi n/J - 2)\hat{u}_{m,n} = h^2 \hat{f}_{m,n} \qquad [3.32]$$

when the variables $x_0, \ldots, x_I$ and $y_0, \ldots, y_J$ have been discretized. The goal of the method is to compute $\hat{f}_{m,n}$ and $\hat{u}_{m,n}$.

The value of $\hat{f}_{m,n}$ is found using the equation:

$$\hat{f}_{m,n} = \sum_{l=0}^{I-1} \sum_{k=0}^{J-1} e^{2i\pi lm/I} e^{2i\pi nk/J} f_{l,k}. \quad [3.33]$$

The value of $\hat{u}_{m,n}$ is then computed using the discretized equation:

$$\hat{u}_{m,n} = \frac{h^2 \hat{f}_{m,n}}{2(\cos 2\pi m/I + \cos 2\pi n/J - 2)}. \quad [3.34]$$

Finally, $u_{i,j}$ is calculated using the inversion formula:

$$u_{l,k} = \frac{1}{IJ} \sum_{m=0}^{I-1} \sum_{n=0}^{J-1} e^{-2i\pi lm/I} e^{-2i\pi nk/J} \hat{u}_{m,n}. \quad [3.35]$$

3.5. Parabolic equations

3.5.1. *Heat equation*

3.5.1.1. *Theta-scheme method*

The heat equation

$$\frac{\partial u}{\partial t} = c \frac{\partial^2 u}{\partial x^2} \quad [3.36]$$

is discretized by

$$\frac{u_{i,j+1} - u_{i,j}}{\Delta t} = \theta c \frac{u_{i-1,j+1} - 2u_{i,j+1} + u_{i+1,j+1}}{(\Delta x)^2} + (1-\theta) c \frac{u_{i-1,j} - 2u_{i,j} + u_{i+1,j}}{(\Delta x)^2}. \quad [3.37]$$

The θ-scheme is known as the explicit method when $\theta = 0$, the Crank–Nicholson method when $\theta = 1/2$, and the implicit method when $\theta = 1$.

For $0 \leq \theta < \frac{1}{2}$, it can be shown that the scheme is stable whenever

$$\frac{c\Delta t}{(\Delta x)^2} \leq \frac{1}{2(1-2\theta)}. \quad [3.38]$$

For $\frac{1}{2} \leq \theta \leq 1$, the method is universally stable.

3.5.1.2. *Alternating direction implicit method (Peaceman–Rachford–Douglas)*

To solve equations of the form

$$\frac{\partial u}{\partial t} = a\frac{\partial^2 u}{\partial x^2} + b\frac{\partial^2 u}{\partial y^2}, \qquad [3.39]$$

Peaceman–Rachford suggested replacing the differential equation by an expression that alternates between two discretized equations after each period $\frac{\Delta t}{2}$. Writing $v_{i,j}$ for the intermediate result, this scheme can be stated as follows:

$$\begin{cases} \dfrac{v_{i,j} - u^n_{i,j}}{(\Delta t/2)} = a\dfrac{v_{i-1,j} - 2v_{i,j} + v_{i+1,j}}{(\Delta x)^2} + b\dfrac{u^n_{i,j-1} - 2u^n_{i,j} + u^n_{i,j+1}}{(\Delta y)^2} \\[2ex] \dfrac{u^{n+1}_{i,j} - v_{i,j}}{(\Delta t/2)} = a\dfrac{v_{i-1,j} - 2v_{i,j} + v_{i+1,j}}{(\Delta x)^2} + b\dfrac{u^{n+1}_{i,j-1} - 2u^{n+1}_{i,j} + u^{n+1}_{i,j+1}}{(\Delta y)^2} \end{cases} \qquad [3.40]$$

Setting

$$\mathcal{L}_x u^n_{i,j} = \frac{u^n_{i-1,j} - 2u^n_{i,j} + u^n_{i+1,j}}{(\Delta x)^2} \qquad [3.41]$$

and

$$\mathcal{L}_y u^n_{i,j} = \frac{u^n_{i,j-1} - 2u^n_{i,j} + u^n_{i,j+1}}{(\Delta y)^2}, \qquad [3.42]$$

and defining U_n and V_n to be the matrices $(u^n_{i,j})$ and $(v^n_{i,j})$, these equations can be written in matrix form:

$$\begin{cases} V_n - U_n = \frac{\Delta t}{2}(a\mathcal{L}_x V_n + b\mathcal{L}_y U_n) \\ U_{n+1} - V_n = \frac{\Delta t}{2}(a\mathcal{L}_x V_n + b\mathcal{L}_y U_{n+1}) \end{cases} \qquad [3.43]$$

At the n-th step, with knowledge of U_n, we can compute V_n, then U_{n+1}, by solving tridiagonal systems.

3.6. Using *Matlab*

This example is taken from the *Matlab* documentation. It solves the Laplace equation on an L-shaped domain. The script is as follows:

```
R = 'L'; % Specify the shape of the domain
% Generate and display the grid
n = 32;
G = numgrid(R,n);
spy(G)
title('A finite difference grid')
% Show a smaller version as sample
g = numgrid(R,12)
D = delsq(G);
spy(D)

title('The 5-point Laplacian')

% Number of interior points
N = sum(G(:)>0)
rhs = ones(N,1);
if (R == 'N')     % Specify boundary conditions
    spparms('autommd',0)
    u = D\rhs;
    spparms('autommd',1)
else
    u = D\rhs; % Useful in the case R=='L'
end
% Draw the contours of the solution
U = G;
U(G>0) = full(u(G(G>0)));
clabel(contour(U)); prism axis
square ij
colormap((cool+1)/2);

mesh(U)
axis([0 n 0 n 0 max(max(U))])
axis
square ij
```

Running this script returns the following results:

```
g =

     0     0     0     0     0     0     0     0     0     0     0     0
     0     1     6    11    16    21    26    36    46    56    66     0
     0     2     7    12    17    22    27    37    47    57    67     0
     0     3     8    13    18    23    28    38    48    58    68     0
     0     4     9    14    19    24    29    39    49    59    69     0
     0     5    10    15    20    25    30    40    50    60    70     0
     0     0     0     0     0     0    31    41    51    61    71     0
     0     0     0     0     0     0    32    42    52    62    72     0
     0     0     0     0     0     0    33    43    53    63    73     0
     0     0     0     0     0     0    34    44    54    64    74     0
     0     0     0     0     0     0    35    45    55    65    75     0
     0     0     0     0     0     0     0     0     0     0     0     0

N =

   675
```

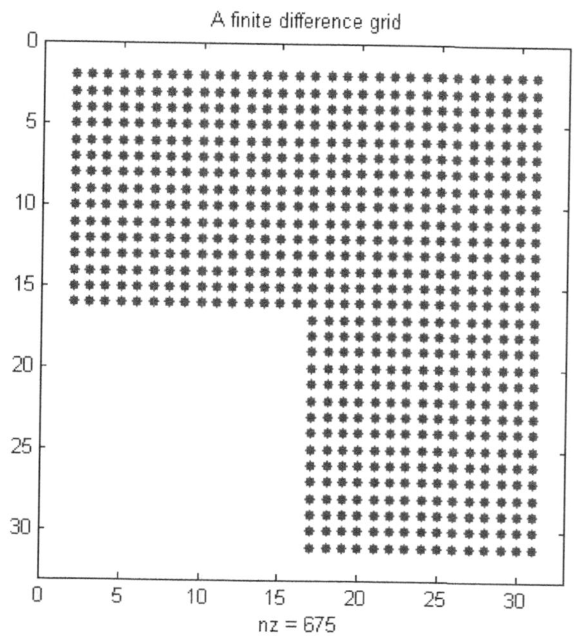

Figure 3.2. *A finite difference grid*

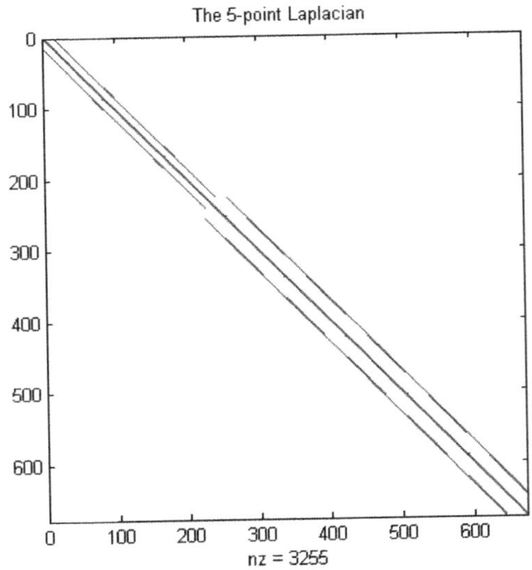

Figure 3.3. *The 5-point Laplacian*

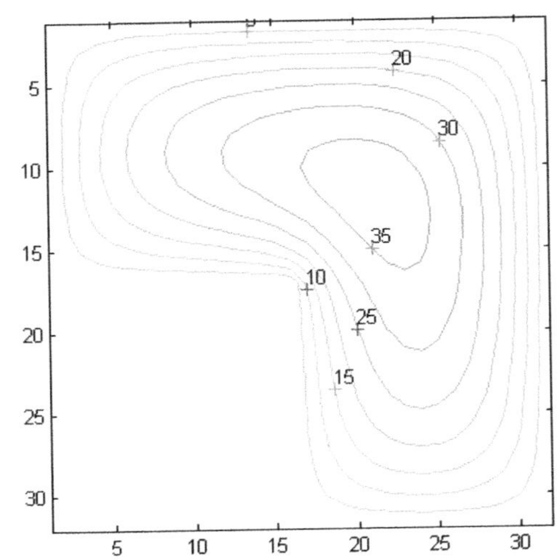

Figure 3.4. *The contours of the solution. For a color version of this figure, see www.iste.co.uk/radi/advanced2.zip*

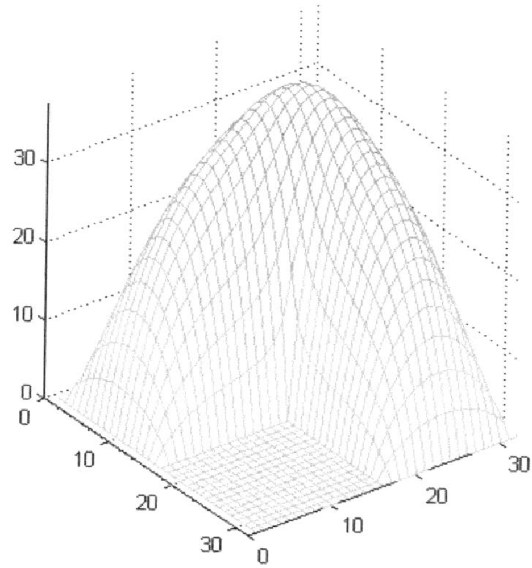

Figure 3.5. *Visualization of the solution. For a color version of this figure, see www.iste.co.uk/radi/advanced2.zip*

Consider the following boundary value problem:

$$-u''(x) + \pi^2 u(x) = \pi^2 x \sin(\pi x) - 2\pi \cos(\pi x) \text{ on }]0, 1[\qquad [3.44]$$

$$u(0) = u(1) = 0 \qquad [3.45]$$

The script listed below can be used to solve this PDE using the method of finite differences [KOK 09]:

```
%-----------------------------------------------------------
% use finite differences to solve the boundary value problem
% -u''(x)=(\pi^2)*x*sin(x*\pi)-2*\pi*cos(x*\pi)
% u(0)=u(1)=0;
%-----------------------------------------------------------
a=0; b=1;
% Subdivision
N=63;
h=(b-a)/(N+1);
x=[a+h:h:b-h]';
% System matrix
e=ones(N,1)/(h*h);
```

```
ee=[-e 2*e -e];
A=pi*pi*speye(N)+spdiags(ee,-1:1,N,N);
% Right-hand side
b=2*pi*pi*x.*sin(pi*x)-2*pi*cos(pi*x);
% Solve
u=A/b;
% Errors
ec=x.*sin(pi*x);
ec1=norm(A*ue-b,inf);          % consistency
ec2=norm(ue-u,inf);            % convergence
fprintf('Consistency error: %15.8e  convergence error: %15.8e
 \n',ec1,ec2)
```

Consider the following problem:

$$\frac{\partial u}{\partial t}(x,t) - \frac{\partial^2 u}{\partial x^2} = 0,\ 0 < x < 1,\ 0 < t \leq T \tag{3.46}$$

$$u(0,t) = u(1,t) = 0,\ 0 < t \leq T \tag{3.47}$$

$$u(x,0) = \sin(\pi x),\ 0 \leq x \leq 1 \tag{3.48}$$

The exact solution of Problem [3.46–3.48] is $u(x) = e^{-\pi^2 t}\sin(\pi x)$, $\forall(x,t) \in [0,1] \times [0,T]$.

The following script can be used to solve this PDE:

```
a=0; b=1;
nu=1; T=0.5;
% Discretization
M=10000; N=1600;
h=(b-a)/(M+1); k=T/N;
theta=0.75; x=[a+h:h:b-h]';
% System matrix
e=ones(M,1);
ee=[-e 2*e -e];
R=spdiages(ee,-1:1,M,M);
A=(1/k)*speye(M)+(theta*nu/h/h)*R;
B=(1/k)*speye(M)-((1-theta)*nu/h/h)*R;

% Right-hand side is constant in time/space
b=zeros(M,1);
```

```
% Time loop
ec=zeros(N,1);
up=sin(pi*x);
it=0;
while(it<N)
    it=it+1;
    t=it+k

% Solve the system
    bt=B*up;

% Solve the system
    ut=A\bt;
    up=ut;
% Error ||u-\pi(u)||_2
    ue=sin(pi*x)*exp(-pi*pi*t);
    ec(it)=sqrt(sum(h*(ut-ue).*(ut-ue)));
end

% Convergence error
ecn=max(ec);
fprintf('Convergence error: %15.8e \n',ecn)
```

4

Finite Element Method

4.1. Introduction

The objective of this chapter is to present the basic principles of the finite element method, with emphasis on how the consistency of the computational process is guaranteed at each step of the method. Analyzing and interpreting the results of a computation requires a solid understanding of each mathematical step in the approximation; only then can the error between the numerical model and the exact solution of the mathematical problem be estimated. We also need to bear in mind that the results of the numerical model can only provide insight into aspects of the mathematical model that are captured by the modeling hypotheses [FIS 07, HAR 07]. In this chapter, we will focus on examples of elementary models that are used for linear theories. These models are already sufficiently rich to solve a wide range of engineering problems.

4.2. One-dimensional finite element methods

Consider the following problem (P):

$$-u''(x) + c(x)u(x) = f(x) \text{ for } 0 \leq x \leq 1 \qquad [4.1]$$

$$u(0) = u(1) = 0 \qquad [4.2]$$

where $u(x)$ is twice continuously differentiable on $[0, 1]$. The function $c(x)$ is assumed to be continuous and positive on $[0, 1]$.

Let V be the vector space of all piecewise continuously differentiable functions $v(x)$ on $[0,1]$ that satisfy the condition $v(0) = v(1)$. Consider an arbitrary element $v(x)$ of V. After multiplying [4.1] by $v(x)$, we can establish the following relation:

$$-\int_0^1 u''(x)v(x)dx + \int_0^1 c(x)u(x)v(x)dx = \int_0^1 f(x)v(x)dx. \qquad [4.3]$$

We can integrate the first term by parts over $[0,1]$ to obtain:

$$-[u'(x)v(x)]_0^1 + \int_0^1 u'(x)v'(x)dx + \int_0^1 c(x)u(x)v(x)dx = \int_0^1 f(x)v(x)dx. \quad [4.4]$$

Since $v(0) = v(1) = 0$, this implies that:

$$\int_0^1 u'(x)v'(x)dx + \int_0^1 c(x)u(x)v(x)dx = \int_0^1 f(x)v(x)dx. \qquad [4.5]$$

Therefore, for any given $v(x)$ in V, every solution $u(x)$ of (P) satisfies equation [4.5].

The problem (P) therefore implies the following problem (PF):

Find a twice continuously differentiable function u on $[0,1]$ with the following properties:

– for every function $v(x)$ in V,

$$\int_0^1 u'(x)v'(x)dx + \int_0^1 c(x)u(x)v(x)dx = \int_0^1 f(x)v(x)dx; \qquad [4.6]$$

– $u(0) = u(1) = 0$.

In fact, it is possible to show that the problems (P) and (PF) both have a unique solution, which implies that they are equivalent. The formulation (PF) is known as the weak formulation.

The advantage of the weak formulation (PF) is that it is linear in $v(x)$. Let $\phi_1, \phi_2, \ldots, \phi_N$ be N linearly independent functions in V. In practice, these functions will depend on a parameter $h > 0$ that we will introduce later.

Write V_h for the vector space generated by $\phi_1, \phi_2, \ldots, \phi_N$. If $w(x)$ is an element of V_h, then:

$$w(x) = \sum_{j=1}^{N} \lambda_j \phi_j(x).$$

Note that $w(0) = w(1) = 0$.

The N-dimensional vector space V_h does not span the full space V (which is infinite-dimensional), but, as N increases, it becomes an increasingly large subspace of V.

The idea of Galerkin approximation is to find u_h in V_h that satisfies a weaker version of the problem, which is obtained by relaxing the requirement that the relation should hold for every $v(x)$ in V, instead only requiring it to hold for every $v_h(x)$ in V_h. Thus, the Galerkin approximation (PF$_h$) of the problem (P) can be stated as follows:

Find u_h in V_h satisfying

$$\int_0^1 u_h'(x)v_h'(x)dx + \int_0^1 c(x)u_h(x)v_h(x)dx = \int_0^1 f(x)v_h(x)dx \qquad [4.7]$$

for every function $v_h(x)$ in V_h.

The Galerkin approximation is weaker than both (P) and (PF). It represents an "approximation" of these problems that improves as the space V_h covers more of the full space V, that is, as N increases.

REMARK.– Every function u_h in V_h is of the form

$$u_h(x) = \sum_{k=0}^{N} \lambda_k \phi_k(x).$$

Finding u_h is therefore equivalent to finding the coefficients $\lambda_1, \lambda_2, \ldots, \lambda_N$.

In order for a given function to be a solution of the problem (PF$_h$), it simply needs to satisfy [4.7] when $v_h(x)$ is taken to be equal to each of the functions $\phi_1, \phi_2, \ldots, \phi_N$. Hence, we can reduce the problem (PF$_h$) to a system of N equations in the N unknowns $\lambda_1, \lambda_2, \ldots, \lambda_N$:

$$\sum_{k=0}^{N} \lambda_k \int_0^1 (\phi_k'(x)\phi_j'(x) + c(x)\phi_j(x)\phi_k(x))dx = \int_0^1 f(x)\phi_j(x)dx.$$

For $j = 1, 2, \ldots, N$, this system can be written as:

$$A \begin{pmatrix} \lambda_1 \\ \lambda_2 \\ \vdots \\ \lambda_N \end{pmatrix} = \begin{pmatrix} F_1 \\ F_2 \\ \vdots \\ F_N \end{pmatrix}, \qquad [4.8]$$

where

$$F_k = \int_0^1 f(x)\phi_k(x)dx, \quad [4.9]$$

and the element A_{jk} of the matrix A is given by

$$A_{jk} = \int_0^1 \phi'_j(x)\phi'_k(x)dx + \int_0^1 c(x)\phi_j(x)\phi_k(x)dx. \quad [4.10]$$

To fully take advantage of the finite element method, we need to choose the N linearly independent functions $\phi_1, \phi_2, \ldots, \phi_N$ carefully. We must consider the following two factors:

– The space V_h needs to cover as much of the original space V as possible. In other words, the function $u_h(x)$ needs to be a good approximation of the original target function $u(x)$.

– The matrix A needs to be as simple as possible, since we will need to solve a very large system of equations; in practice, we need to choose the functions $\phi_j(x)$ in such a way that as many elements as possible of the matrix A are zero.

One classical and straightforward way of choosing $\phi_1, \phi_2, \ldots, \phi_N$ is as follows.

First, define $h = \frac{1}{N+1}$, and discretize the interval $[0, 1]$ by setting $x_k = kh$ for $k = 0, 1, 2, \ldots, N+1$.

The x_k are the nodes of the discretization. Now, for $k = 0, 1, 2, \ldots, N+1$, consider the function $\phi_k(x)$ shown in Figure 4.1.

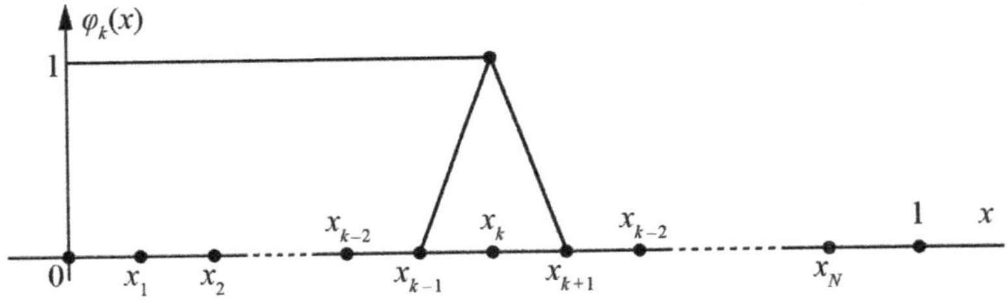

Figure 4.1. *The function ϕ*

Setting

$$B_{jk} = \int_0^1 \phi'_j(x)\phi'_k(x)dx \quad \text{and} \quad C_{jk} = \int_0^1 c(x)\phi_j(x)\phi_k(x)dx, \quad [4.11]$$

We can rewrite equation [4.10] as $A_{jk} = B_{jk} + C_{jk}$. The next step is to evaluate B_{jk}, C_{jk}, and F_k. The first thing to note is that:

$$\phi'_j(x) = \begin{cases} 0 & \text{if } x < x_{j-1} \text{ or } x > x_{j+1} \\ \frac{1}{h} & \text{if } x_{j-1} < x < x_j \\ -\frac{1}{h} & \text{if } x_j < x < x_{j+1} \end{cases} \qquad [4.12]$$

Therefore, B_{jk} is given by

$$B_{jk} = \begin{cases} \frac{2}{h} & \text{if } j = k \\ -\frac{1}{h} & \text{if } j = k \pm 1 \\ 0 & \text{otherwise} \end{cases} \qquad [4.13]$$

We can use the trapezoidal rule with a step size of h to evaluate the integrals C_{jk} and F_k. This gives us an approximate value for $\int_0^1 h(x)dx$:

$$\frac{1}{2}h(x_0) + \sum_{j=1}^{N} h(x_j) + \frac{1}{2}h(x_{N+1}).$$

When $h(0) = h(1) = 0$, the above formula reduces to $\sum_{j=1}^{N} h(x_j)$. In our case,

$$C_{jk} = \begin{cases} hc(x_j) & \text{if } j = k \\ 0 & \text{if } j \neq k \end{cases}, \quad F_k = hf(x_k). \qquad [4.14]$$

After defining $c_k = c(x_k)$ and $f_k = f(x_k)$, the matrix A can be expressed as the following tridiagonal matrix:

$$\frac{1}{h}\begin{pmatrix} 2+h^2c_1 & -1 & 0 & \cdots & 0 & 0 & 0 \\ -1 & 2+h^2c_2 & 1 & \cdots & 0 & 0 & 0 \\ 0 & -1 & 2+h^2c_3 & -1 & \cdots & 0 & 0 \\ 0 & 0 & -1 & \cdots & 0 & 0 & 0 \\ 0 & 0 & 0 & \cdots & 0 & 0 & 0 \\ \vdots & \vdots & \vdots & \vdots & \vdots & \vdots & \vdots \\ 0 & 0 & 0 & \cdots & -1 & 2+h^2c_{N-1} & -1 \\ 0 & 0 & 0 & \cdots & 0 & -1 & 2+h^2c_N \end{pmatrix}. \qquad [4.15]$$

The system that we need to solve is

$$A \begin{pmatrix} \lambda_1 \\ \lambda_2 \\ \vdots \\ \lambda_N \end{pmatrix} = h \begin{pmatrix} f_1 \\ f_2 \\ \vdots \\ f_N \end{pmatrix}. \qquad [4.16]$$

By writing $u(x)$ for the exact solution of the problem (P) as before, and $u_h(x)$ for the solution obtained by solving the system [4.16], it is possible to show that

$$|u(x) - u_h(x)| \leq Ch^2, \qquad [4.17]$$

where C is a constant. Hence, this method is $O(h^2)$ in these conditions, which is already very good.

4.3. Two-dimensional finite element methods

Let Ω be a polygonal contour, and suppose that $F(x, y)$ is continuously differentiable on Ω (see Figure 4.2).

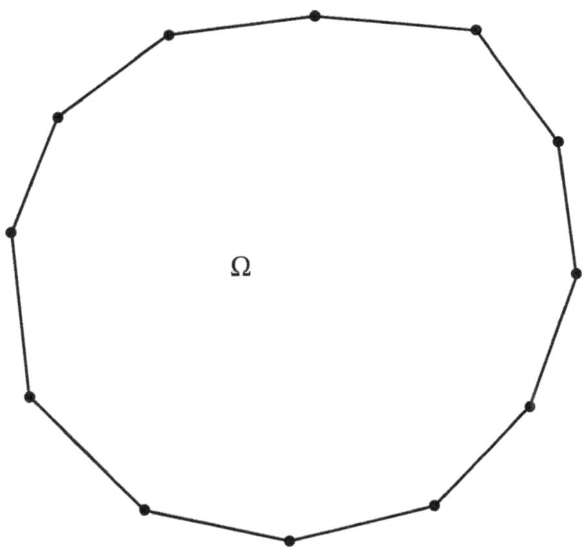

Figure 4.2. *The domain* Ω

Consider the following problem (P):

Find a twice continuously differentiable function $u(x, y)$ on Ω that satisfies:

$$-\Delta u = F(x, y) \text{ for all } (x, y) \in \Omega; \quad [4.18]$$

$$u(x, y) = 0 \text{ for all } (x, y) \text{ on the boundary of } \Omega. \quad [4.19]$$

Let $H_0^1(\Omega)$ be the vector space of piecewise continuously differentiable functions on Ω that vanish on the boundary of Ω.

Suppose that $u(x, y)$ is a solution of the problem (P). Let $v(x, y)$ be an element of $H_0^1(\Omega)$. Multiplying [4.18] by $v(x, y)$, then integrating over Ω yields:

$$-\int\int_\Omega (\Delta u) v dx dy = \int\int_\Omega F v dx dy. \quad [4.20]$$

We know that

$$\int\int_\Omega (\Delta u) v dx dy = \int\int_\Omega (\frac{\partial^2 u}{\partial x^2} + \frac{\partial^2 u}{\partial y^2}) v dx dy. \quad [4.21]$$

By Stokes' theorem,

$$\int\int_\Omega (\Delta u) v dx dy = -\int\int_\Omega (\nabla u)(\nabla v) dx dy + \int_{\partial\Omega} \frac{\partial u}{\partial n} v dn. \quad [4.22]$$

However, v vanishes on the boundary, so

$$\int\int_\Omega (\Delta u) v dx dy = -\int\int_\Omega \nabla u . \nabla v dx dy. \quad [4.23]$$

Therefore, $u(x, y)$ satisfies

$$\int\int_\Omega \nabla u . \nabla v dx dy = \int\int_\Omega F v dx dy. \quad [4.24]$$

This reasoning can be reversed, so the problem (P) is equivalent to the following problem (PF):

Find a twice continuously differentiable function $u(x, y)$ on Ω that satisfies:

– for every function $v(x, y)$ in $H_0^1(\Omega)$,

$$\int\int_\Omega \nabla u . \nabla v dx dy = \int\int_\Omega F v dx dy;$$

– for each point (x, y) on the boundary of Ω, $u(x, y) = 0$.

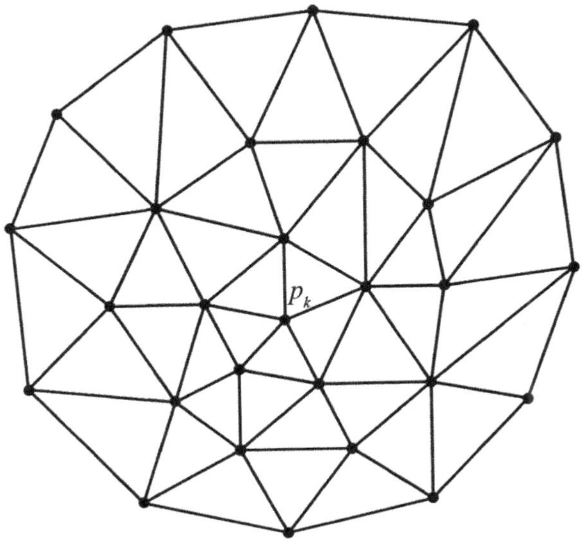

Figure 4.3. *Triangulation*

Next, consider a triangulation T_h of the polygonal contour Ω:

Given any pair of triangles in this triangulation, one of the following conditions holds that:

– both triangles are disjoint;

– the triangles share precisely one vertex;

– the triangles share precisely one edge.

Here, we take h to be the maximum diameter of the triangles in the triangulation.

The nodes are the vertices of the triangles that are not on the boundary of Ω. Let $p_1, p_2, \ldots, p_k, \ldots, p_N$ be an enumeration of these nodes. For each node p_k, consider the function $\phi_k(x, y)$ whose graph is shown in Figure 4.4.

– Consider the polygon B_k formed by the points of the triangulation that are directly connected to the node p_k.

– Outside of B_k, the function $\phi_k(x, y)$ is 0.

– Next, construct the pyramid with base B_k whose summit is height 1 vertically above the node p_k.

– On B_k, the graph of $\phi_k(x, y)$ is defined as the faces of the pyramid. In other words, the points $(x, y, \phi_k(x, y))$ sweep over the faces of the pyramid; in particular, $\phi_k(p_k) = 1$ and $\phi_k(x, y)$ is 0 for every point (x, y) on the contour of B_k.

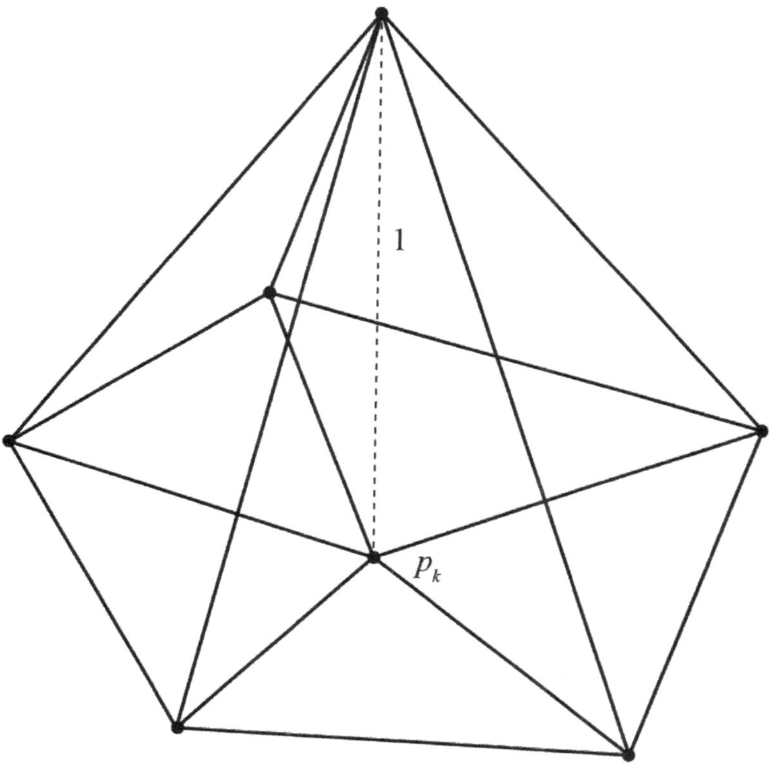

Figure 4.4. *Graph*

By construction, each function $\phi_k(x, y)$ vanishes at every point on the boundary of Ω.

Now, consider the vector space V_h of linear combinations of the functions $\phi_1, \phi_2, \ldots, \phi_N$. Every element w of V_h is of the form

$$w(x, y) = \sum_{k=1}^{N} \lambda_k \phi_k(x, y). \qquad [4.25]$$

The vector space V_h is an approximation of $H_0^1(\Omega)$. The quality of this approximation naturally depends on the triangulation, and in particular the number of nodes.

The problem (PF) can therefore be approximated by the following weaker problem, (PF$_h$), known as the Galerkin approximation of the original problem:

Find $u_h(x, y)$ in V_h such that

$$\int\int_\Omega \nabla u_h.\nabla v_h dxdy = \int\int_\Omega F v_h dxdy \qquad [4.26]$$

for every function $v_h(x, y)$ in V_h.

As the problem (PF) is linear in $v(x)$, equation [4.26] is also linear in $v_h(x)$. Therefore, for a function to satisfy (PF$_h$), it simply needs to satisfy [4.26], when $v_h(x)$ is equal to each of the functions $\phi_1(x, y), \phi_2(x, y), \ldots, \phi_N(x, y)$.

As before, to determine u_h, we simply need to find $\lambda_1, \lambda_2, \ldots, \lambda_N$ such that

$$u_h(x, y) = \sum_{j=1}^{N} \lambda_k \phi_k(x, y) \qquad [4.27]$$

is a solution of [4.26].

Thus, we solve the following system of N equations in the N unknowns $\lambda_1, \lambda_2, \ldots, \lambda_N$:

$$\sum_{k=0}^{N} \lambda_k \int\int_\Omega \nabla\phi_j.\nabla\phi_k dxdy = \int\int_\Omega F\phi_j dxdx \quad \text{for } j = 1, 2, \ldots, N. \qquad [4.28]$$

This system can be written in matrix form as:

$$A \begin{pmatrix} \lambda_1 \\ \lambda_2 \\ \vdots \\ \lambda_N \end{pmatrix} = \begin{pmatrix} F_1 \\ F_2 \\ \vdots \\ F_N \end{pmatrix}, \qquad [4.29]$$

where

$$F_j = \int\int_\Omega F\phi_j dxdy. \qquad [4.30]$$

The element A_{jk} of the matrix A is given by

$$A_{jk} = \int\int_{\Omega} (\nabla\phi_j).(\nabla\phi_k)dxdy. \quad [4.31]$$

Many of the terms A_{jk} will be zero, since each function $\phi_k(x, y)$ is zero outside of the polygon B_k, hence the term A_{jk} is zero whenever the node p_j is not directly connected to the node p_k.

4.4. General procedure of the method

The process of constructing a finite element model can be divided into the following key steps [BAT 96]:

– discretize the continuous medium into subdomains;

– construct a nodal approximation on each subdomain;

– compute the elementary matrices of the integral form of the problem;

– assemble the elementary matrices;

– account for the boundary conditions;

– solve the system of equations.

4.5. Finite element method for computing elastic structures

4.5.1. *Linear elasticity*

This section briefly reviews a few notions from linear elasticity to ensure that the model problem is clearly stated, and introduces the notation that we will use below. We will study the static equilibrium of a deformable solid body S whose undeformed state occupies a bounded domain Ω in the affine space $\mathbb{R}^d$ (usually $d = 1, 2,$ or 3) (Figure 4.5). The space $\mathbb{R}^d$ is equipped with an affine orthonormal coordinate system $\{O, e_1, \ldots, e_d\}$. The coordinates of a point $x \in \mathbb{R}^d$ are denoted $(x_1, \ldots, x_d)$. Each vector u in the underlying vector space has components $(u_1, \ldots, u_d)$ with respect to the orthonormal basis $\{e_1, \ldots, e_d\}$.

The body S is subjected to a stress f (per unit volume), as well as a traction or compression φ_N (per unit surface area) on a region of its boundary Γ_N, and an imposed displacement φ_D on the rest of its boundary Γ_D. We make the following assumptions:

$$\Gamma_D \cup \Gamma_N = \Gamma, \ \Gamma_D \cap \Gamma_N = \emptyset, \text{ and } \Gamma_D \neq \emptyset. \quad [4.32]$$

The third condition in [4.32] is required to eliminate any rigid body motion. Under the action of the external stresses, the body S deforms to occupy a new domain Ω_d in $\mathbb{R}^d$, which we shall call the updated or deformed configuration.

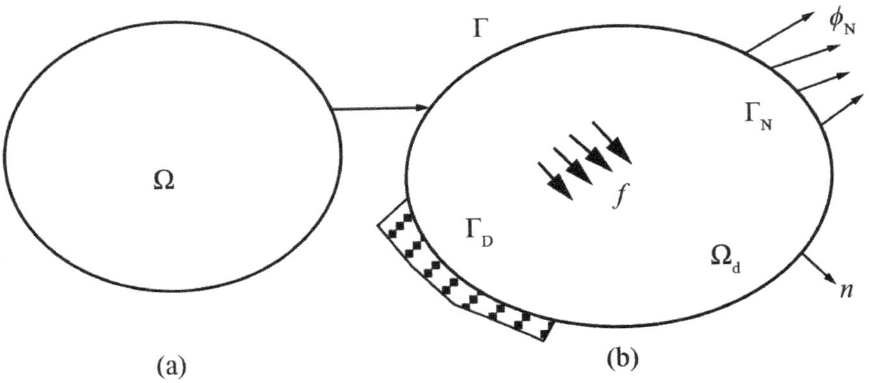

Figure 4.5. *Elastic body in equilibrium under external loads: a) initial configuration of S; b) deformed configuration of S*

If the small perturbation hypothesis (SPH) holds, the original configuration Ω and the deformed configuration Ω_d of S are assumed to be identical. This allows us to write the static equilibrium equations on Ω:

$$\begin{aligned} &\text{div}\sigma(x) + f(x) = 0 && \forall x \in \Omega \text{ (by the balance of forces)} \\ &\sigma(x).n(x) = \varphi_N && \forall x \in \Gamma_N \\ &\sigma(x) = (\sigma(x))^t && \forall x \in \Omega \text{ (by the balance of moments)} \end{aligned} \qquad [4.33]$$

where $\sigma = (\sigma_{ij})_{i,j=1,\dots,d}$ denotes the second-order Cauchy stress tensor, σ^t denotes its transpose, and the divergence operator is defined by $\text{div}\sigma(x) = \text{tr}(D_x\sigma(x))$, or in other words $(\text{div}\sigma(x))_i = \partial_{x_j}\sigma_{ij}(x)$ for $i = 1, \dots, d$, in summation notation. In the following, we will no longer distinguish between the body S itself and the domain Ω that it occupies.

As applying the same external load to different materials leads to different deformations, the equilibrium equations [4.33] do not fully determine the equilibrium state of the deformable body S. In order for the problem to be fully determined, we need to consider the behavioral law of the material from which S is made. According to the SPH, the (second-order) tensor of linearized deformations

$\varepsilon(x) = (\varepsilon_{ij})_{i,j=1,\dots,d}$ satisfies the following linear relation with the displacement field $u\colon \Omega \to \mathbb{R}^d$:

$$\varepsilon(u) = \frac{1}{2}(D_x u + D_x^t u), \quad \text{or, in summation notation,}$$
$$\varepsilon_{ij}(u) = \frac{1}{2}(\partial_{x_j} u_i + \partial_{x_i} u_j). \qquad [4.34]$$

If we further assume that Ω is made from a linear elastic material, then its behavioral law (or constitutive equations), which relate the Cauchy stress tensor σ to the strain tensor ε, is simply Hooke's law:

$$\sigma(x) = E(x)\varepsilon(x), \quad \text{or, in summation notation,}$$
$$\sigma_{ij}(x) = E_{ijkl}(x)\varepsilon_{kl}(x). \qquad [4.35]$$

The (fourth-order) elasticity tensor E has the following symmetry properties:

$$E_{ijkl} = E_{klij} \quad \text{(major symmetry)}$$
$$E_{ijkl} = E_{jikl} = E_{jilk} \quad \text{(minor symmetry)} \qquad [4.36]$$

Note that the major symmetry follows from elasticity and the minor symmetry follows from the symmetry of stresses and strains. Moreover, E is defined positive:

$$E_{ijkl}(x)\psi_{ij}\psi_{kl} \geq 0$$
$$E_{ijkl}(x)\psi_{ij}\psi_{kl} = 0 \Rightarrow \psi_{ij} = 0, \ \forall x, \forall \psi \text{ such that } \psi_{ij} = \psi_{ji}. \qquad [4.37]$$

In homogeneous materials, E is independent of x. In isotropic materials, E is uniquely determined by two strictly positive constants, λ and μ, known as the Lamé coefficients. These coefficients satisfy the following relation with Young's modulus η and the Poisson coefficient ν:

$$E_{ijkl} = \lambda\delta_{ij}\delta_{kl} + \mu(\delta_{il}\delta_{jk} + \delta_{ik}\delta_{jl})$$
$$\lambda = \frac{\eta\nu}{(1+\nu)(1-2\nu)}, \ \mu = \frac{\eta}{2(1+\nu)}, \qquad [4.38]$$

where δ_{ij} is the Kronecker delta ($\delta_{ij} = 1$ if $i = j$, $\delta_{ij} = 0$ if $i \neq j$). In this case, the behavior relation [4.35] can be rewritten as follows:

$$\begin{aligned}
\sigma(x) &= \lambda \text{tr}(\varepsilon(x))I + 2\mu\varepsilon(x) \\
\varepsilon(x) &= \frac{1+\nu}{\eta}\sigma(x) - \frac{\nu}{\eta}\text{tr}(\sigma(x))I, \text{ or, in summation notation,} \\
\sigma_{ij}(x) &= \lambda\varepsilon_{kk}(x)\delta_{ij} + 2\mu\varepsilon_{ij}(x) \\
\varepsilon_{ij}(x) &= \frac{1+\nu}{\eta}\sigma_{ij}(x) - \frac{\nu}{\eta}\sigma_{kk}(x)\delta_{ij}.
\end{aligned} \qquad [4.39]$$

Finally, by choosing the displacement field u as the primary unknown of the problem and grouping together the equilibrium equations [4.33], the definition of the linearized deformations [4.34] and the behavioral law [4.39], we obtain the *strong* or *local* equilibrium formulation of our static linear elasticity problem:

Given an elastic body $\Omega \subset \mathbb{R}^d$ whose deformations are defined by [4.34] and whose behavior is governed by the relation [4.39], $f\colon \Omega \to \mathbb{R}^d, \varphi_N\colon \Gamma_N \to \mathbb{R}^d$ and $\varphi_D\colon \Gamma_D \to \mathbb{R}^d$.

Find $u\colon \Omega \to \mathbb{R}^d$, such that:

$$\begin{aligned}
&\text{div}\sigma(x) + f(x) = 0 && \forall x \in \Omega \\
&u = \varphi_D && \text{on} \quad \Gamma_D \\
&\sigma(x).n(x) = \varphi_N && \forall x \in \Gamma_N
\end{aligned} \qquad [4.40]$$

After introducing a second-order differential operator, we can reformulate the linear elasticity problem [4.40] as follows [DUV 98]:

Find $u\colon \Omega \to \mathbb{R}^d$ such that:

$$\begin{cases}
(Lu)_i = f_i & \text{on } \Omega, \quad i = 1, \ldots, d \\
u = \varphi_D & \text{on } \Gamma_D \\
\displaystyle\sum_{j=1}^{d} \sigma_{ij}(u)n_j = (\varphi_N)_i & \text{on } \Gamma_N,\ i = 1, \ldots, d.
\end{cases} \qquad [4.41]$$

The differential operator L is defined as $Lw = -\text{div}\sigma(w)$, and $(Lw)_i = -\sum_{j=1}^{d} D_j\sigma_{ij}(w) = -\sum_{j=1}^{d} \partial_{x_j}\sigma_{ij}(w)$ for $i = 1, \ldots, d$ (no summation notation), where $\lambda \geq 0$, $\mu > 0$ are the Lamé coefficients of the elastic material.

Each of the fields used in this formulation is assumed to be sufficiently regular so that its derivatives exist and the mechanical problem is well-defined. We also assume that the boundary $\Gamma = \partial\Omega$ of the field Ω is sufficiently regular in the sense that the outward unit normal vector $n(x) = (n_j)_{1 \leq j \leq d}$ can be defined almost everywhere. For homogeneous and isotropic materials, the Navier equations hold:

$$\mu\Delta u + (\lambda + \mu)\text{grad}(\text{div}\, u) + f = 0 \quad \text{on } \Omega, \tag{4.42}$$

that is,

$$Lw = -\text{div}\sigma(w) = -\mu\Delta w - (\lambda + \mu)\nabla\text{div}\, w,$$

where $(\Delta u)_i = \sum_k (\frac{\partial^2 u_i}{\partial x_k^2})$ is the Laplace operator, $\text{div}(u) = \frac{\partial u_k}{\partial x_k}$ is the divergence operator and $(\nabla\phi)_i = (\text{grad}\,\phi)_i = \frac{\partial\phi}{\partial x_i}$ is the gradient operator applied to a scalar function ϕ. These equations show that the elasticity problem is described by an elliptic system of second-order partial differential equations. This system can be solved analytically and approximated by the method of finite differences. However, in practice, the complexity of the boundary conditions and the geometric shapes of the solid bodies often prevent an analytic approach, and the finite element method (FEM) tends to be preferred over the finite difference method, which requires a "structured" domain (that can be meshed with right-angled quadrilaterals).

4.5.2. *Variational formulation of the linear elasticity problem*

For simplicity, assume that $\Gamma_N = \emptyset$ and $\varphi_D = 0$ in Problem [4.41], that is, that the solid is fixed along all of its boundary. This leads to the following boundary value problem:

Find $u : \Omega \to \mathbb{R}^d$ such that:

$$\begin{cases} (Lu)_i = f_i & \text{on } \Omega \ 1 \leq i \leq d \\ u_i = 0 & \text{on } \Gamma \ 1 \leq i \leq d \end{cases} \tag{4.43}$$

where

$$(Lu)_i = -2\mu \sum_{j=1}^{d} \frac{\partial}{\partial x_j}\varepsilon_{ij}(u) - \lambda\frac{\partial}{\partial x_i}\text{div}\,(u).$$

4.5.2.1. *Weak formulation of Problem [4.43]*

To derive a variational formulation (mechanically, this is known as the principle of virtual power) for Problem [4.43], we multiply both sides of the PDE by a test function $v = (v_i)_{1\leq i\leq d} \in \left(H_0^1(\Omega)\right)^d$. The Sobolev spaces that provide the theoretical foundation for the variational formulation of this problem are described in Appendix 2.

Let $Lu = ((Lu)_i)_{1\leq i\leq d}$ and $f = (f_i)_{1\leq i\leq d}$. Then:

$$
\begin{aligned}
(Lu, v)_\Omega &= \int_\Omega \sum_i (Lu)_i v_i d\Omega \\
&= \sum_i \int_\Omega - \sum_j D_j(\sigma_{ij}(u))v_i d\Omega \\
&= -\sum_{i,j} (\int_\Omega D_j(\sigma_{ij}(u))v_i) d\Omega \\
&= -\sum_{i,j} (\int_{\partial\Omega} \sigma_{ij}(u)v_i n_j - \int_\Omega \sigma_{ij}(u) D_j v_i d\Omega) \\
&\quad \text{[Green's formula]} \\
&= \sum_{i,j} \int_\Omega \sigma_{ij}(u) D_j v_i d\Omega \ \ (\text{since } v_i/\partial\Omega = 0) \\
&= \sum_{i,j} \int_\Omega (\mu(D_i u_j + D_j u_i) + \lambda \text{div}(u)\delta_{ij}) D_j v_i d\Omega.
\end{aligned}
$$

However, $\sum_{i,j} D_j v_i = \frac{1}{2}\sum_{i,j}(D_i v_j + D_j v_i)$, and the Kronecker delta satisfies $\sum_{i,j} \delta_{ij} D_j v_i = \sum_j D_j v_j = \sum_i D_i v_i = \text{div}(v)$, which implies that:

$$
(Lu, v)_\Omega = \frac{\mu}{2} \int_\Omega \sum_{i,j} (D_i u_j + D_j u_i)(D_i v_j + D_j v_i) + \lambda \int_\Omega \text{div}(u)\text{div}(v).
$$

Introducing the bilinear form

$$
a(u, v) = \frac{\mu}{2} \int_\Omega \sum_{i,j} (D_i u_j + D_j u_i)(D_i v_j + D_j v_i) + \lambda \int_\Omega \text{div}(u)\text{div}(v)
$$

this now allows us to establish the weak formulation of [4.43]:

$$\text{Find } u \in H_0^1(\Omega)^d \text{ such that } \quad a(u,v) = l(v) \quad \forall v \in H_0^1(\Omega)^d, \qquad [4.44]$$

where

$$a(u,v) = 2\mu \sum_{i,j=1}^{d} \int_\Omega \varepsilon_{ij}(u)\varepsilon_{ij}(v)d\Omega + \lambda \int_\Omega \text{div}(u)\text{div}(v)d\Omega \qquad [4.45]$$

and

$$l(v) = \int_\Omega fvd\Omega.$$

Clearly, the bilinear form $a(.,.)$ (resp. the linear form l) is continuous on $(H_0^1(\Omega))^d \times (H_0^1(\Omega))^d$ (resp. continuous on $(H_0^1(\Omega))^d$). We still need to show that $a(.,.)$ is coercive.

To do this, note that $a(v,v) = \frac{\mu}{2}\int_\Omega \sum_{i,j}(D_iv_j + D_jv_i)^2 + \lambda \int_\Omega (\text{div}(v))^2$. However, $(\text{div}(v))^2 \geq 0$ and $\lambda \geq 0$, so $a(v,v) \geq \frac{\mu}{2}\int_\Omega \sum_{i,j}(D_iv_j + D_jv_i)^2$ $\forall v \in (H_0^1(\Omega))^d$. By Korn's inequality (see Appendix 1):

$$\int_\Omega \sum_{i,j}(D_iv_j + D_jv_i)^2 \geq K \|v\|_{1,\Omega}^2 \quad \forall v \in (H_0^1(\Omega))^d,$$

we deduce that $\forall v \in (H_0^1(\Omega))^d$, $a(v,v) \geq \alpha \|v\|_{1,\Omega}^2$, where $\|.\|_{1,\Omega}$ is the norm on $H^1(\Omega)$, and hence, the bilinear form is indeed coercive. It then follows from the Lax–Milgram theorem [BRE 83] that the weak equation [4.44] has a unique solution $u \in (H_0^1(\Omega))^d$.

4.5.3. *Planar linear elasticity problems*

These problems consider symmetric 2×2 tensors, which have three elements. In vector notation, we can write the tensors as follows [HAR 07]:

$$\sigma = [\sigma_{11}, \sigma_{22}, \sigma_{12}]^t, \quad \epsilon = [\epsilon_{11}, \epsilon_{22}, \gamma_{12}]^t.$$

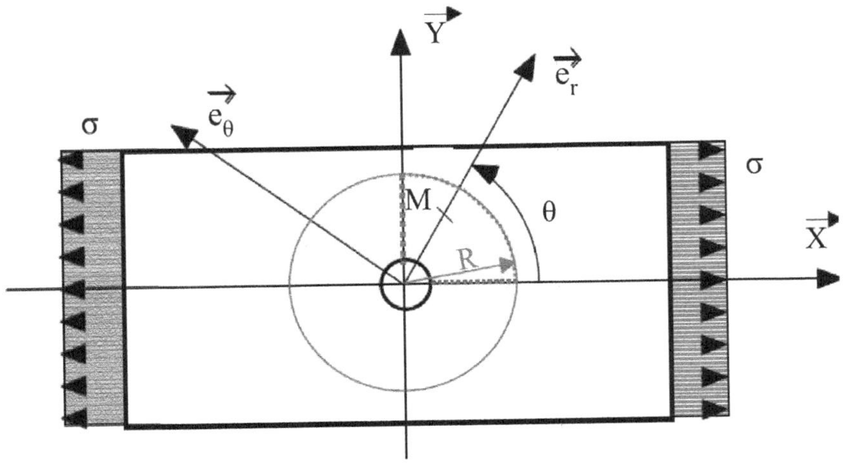

Figure 4.6. *Punctured plate*

There are two types of planar linear elasticity problems:

– Planar stress problems, which assume that $\sigma_{i3} = 0$. For these problems, ϵ_{33} is not zero, but depends on $\epsilon_{11}, \epsilon_{22}$ (since $\sigma_{33} = 0$).

– Planar strain problems, which assume that $\epsilon_{i3} = 0$. For these problems, σ_{33} is not zero, but depends on σ_{11}, σ_{22} (since $\epsilon_{33} = 0$).

When working with isotropic elastic materials, we can deduce the matrix A from Hooke's law:

– Planar stress:

$$A = \frac{E}{1-\nu^2}\begin{pmatrix} 1 & \nu & 0 \\ \nu & 1 & 0 \\ 0 & 0 & \frac{1-\nu}{2} \end{pmatrix}.$$

The value of ϵ_{33} is computed from the relation $\epsilon_{33} = \frac{\nu}{1-\nu}(\epsilon_{11} - \epsilon_{22})$.

– Planar strain:

$$A = \frac{E}{(1+\nu)(1-2\nu)}\begin{pmatrix} 1-\nu & \nu & 0 \\ \nu & 1-\nu & 0 \\ 0 & 0 & \frac{1-2\nu}{2} \end{pmatrix}.$$

The value of σ_{33} is computed from the relation $\sigma_{33} = \nu(\sigma_{11} + \sigma_{22})$.

4.5.4. *Applying the finite element method to planar problems*

For planar problems, we choose triangular or rectangular shapes for the finite elements [CIA 02].

$P1$ triangle with three nodes (see Figure 4.7).

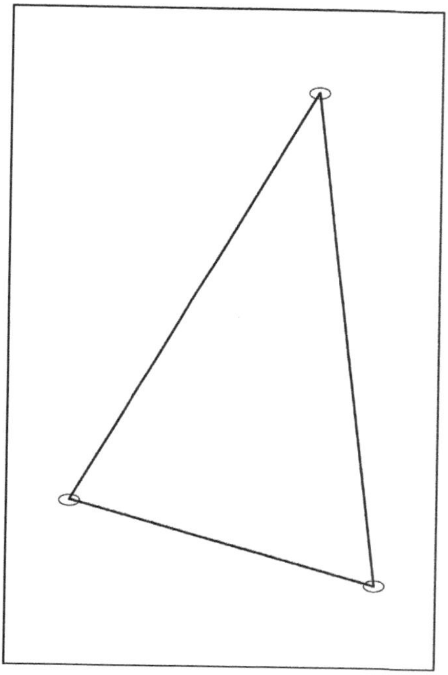

Figure 4.7. *Triangle with three nodes*

The same interpolation nodes are used for both the geometry and the displacements (isoparametric elements) [ZIE 00].

DEFINITION.– *An element is said to be isoparametric if the same interpolations are used to model both the distance and the displacement.*

The three nodes define a polynomial basis ϕ with three terms: $\phi = (1,\ r,\ s)$. The interpolating functions are $N_1 = 1 - r - s$, $N_2 = r$, and $N_3 = s$. The matrix B is computed as follows:

$\epsilon = Bu_n;$

As $\epsilon = (\frac{\partial u}{\partial x}, \frac{\partial v}{\partial y}, \frac{\partial u}{\partial y} + \frac{\partial v}{\partial x})^t$,

$$\begin{pmatrix} \frac{\partial u}{\partial x} \\ \frac{\partial u}{\partial y} \end{pmatrix} = J^{-1} \begin{pmatrix} \frac{\partial u}{\partial r} \\ \frac{\partial u}{\partial s} \end{pmatrix} \quad \text{and} \quad \begin{pmatrix} \frac{\partial v}{\partial x} \\ \frac{\partial v}{\partial y} \end{pmatrix} = J^{-1} \begin{pmatrix} \frac{\partial v}{\partial r} \\ \frac{\partial v}{\partial s} \end{pmatrix} \tag{4.46}$$

$$\begin{array}{l} u = (1-r-s)u_1 + ru_2 + su_3 \\ v = (1-r-s)v_1 + rv_2 + sv_3 \end{array} \quad \text{and} \quad \begin{array}{l} x = (1-r-s)x_1 + rx_2 + sx_3 \\ y = (1-r-s)y_1 + ry_2 + sy_3 \end{array}. \tag{4.47}$$

This gives:

$$\begin{pmatrix} \frac{\partial u}{\partial r} \\ \frac{\partial u}{\partial s} \end{pmatrix} = \begin{pmatrix} -u_1 + u_2 \\ -u_1 + u_3 \end{pmatrix} \quad \text{and} \quad \begin{pmatrix} \frac{\partial v}{\partial r} \\ \frac{\partial v}{\partial s} \end{pmatrix} = \begin{pmatrix} -v_1 + v_2 \\ -v_1 + v_3 \end{pmatrix} \tag{4.48}$$

$$J = \begin{pmatrix} \frac{\partial x}{\partial r} & \frac{\partial y}{\partial r} \\ \frac{\partial x}{\partial s} & \frac{\partial y}{\partial s} \end{pmatrix} = \begin{pmatrix} -x_1 + x_2 & -y_1 + y_2 \\ -x_1 + x_3 & -y_1 + y_3 \end{pmatrix}. \tag{4.49}$$

Hence,

$$J^{-1} = \frac{1}{\Delta} \begin{pmatrix} y_3 - y_1 & y_1 - y_2 \\ x_1 - x_3 & x_2 - x_1 \end{pmatrix}, \tag{4.50}$$

where

$$\Delta = (x_2 - x_1)(y_3 - y_1) - (y_2 - y_1)(x_3 - x_1). \tag{4.51}$$

Thus,

$$\epsilon = \frac{1}{\Delta} \begin{pmatrix} y_2 - y_3 & 0 & y_3 - y_1 & 0 & y_1 - y_2 & 0 \\ 0 & x_3 - x_2 & 0 & x_1 - x_3 & 0 & x_2 - x_1 \\ x_3 - x_2 & y_2 - y_3 & x_1 - x_3 & y_3 - y_1 & x_2 - x_1 & y_1 - y_2 \end{pmatrix} \begin{pmatrix} u_1 \\ v_1 \\ u_2 \\ v_2 \\ u_3 \\ v_3 \end{pmatrix}. \tag{4.52}$$

REMARK.– Note that the matrix B is constant (over the whole element), so the triangle P_1 is implicitly assumed to have constant stress and strain.

By reusing the spatial interpolating functions of the triangle P_1 and adding interpolation nodes for the displacement in the middle of each edge, we can construct a subparametric element that is quadratic in the displacement.

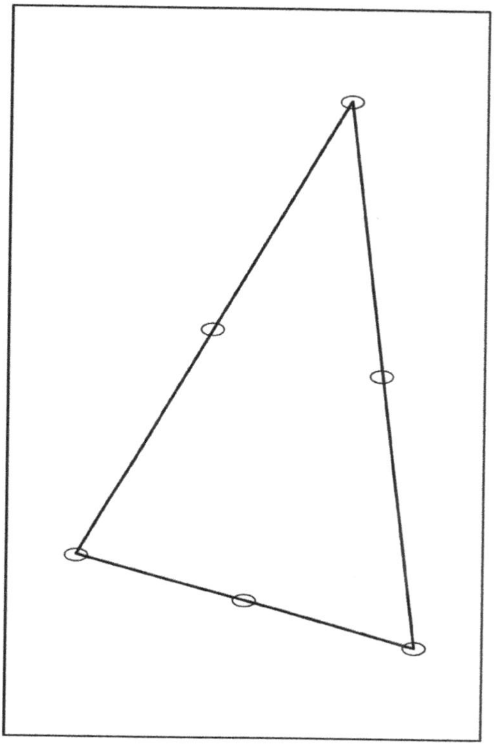

Figure 4.8. *$P2$ triangle*

For isoparametric triangles with 6 nodes (continuity class C^0), we use the same interpolating functions as above, but $N = \bar{N}$. This allows us to define a more realistically curved element with quadratic edges.

For (cubic) triangles with 9 nodes (with continuity class C^0), the functions N are of the third-order; this type of element can also be defined isoparametrically.

For Hermite-type triangles (continuity class C^1), we require the derivative to be continuous at the boundary. There are six degrees of freedom at each of the nodes $1, 2, 3$, namely

$$u, \frac{\partial u}{\partial r}, \frac{\partial u}{\partial s}, v, \frac{\partial v}{\partial r}, \frac{\partial v}{\partial r},$$

and two degrees of freedom at node 4, namely u and v.

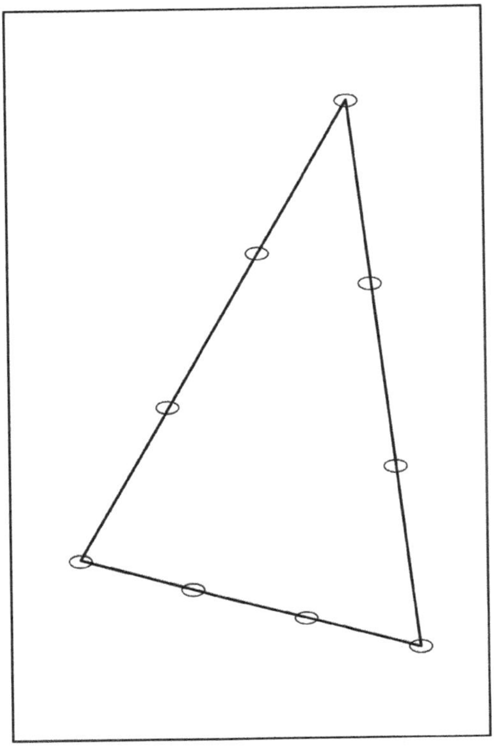

Figure 4.9. *$P3$ triangle*

In one direction, we therefore have a basis ϕ of $(3 \times 3) + 1 = 10$ terms:

$$\phi = (1, r, s, r^2, rs, s^2, r^3, r^2 s, sr^2, s^3). \quad [4.53]$$

The following quadrilateral elements can be found in the literature:

- Q_1 quadrilateral with 4 nodes;
- Q_2 quadrilateral with 9 nodes;
- Serendip quadrilateral with 8 nodes;
- cubic quadrilateral with 16 or 12 nodes;
- Hermite quadrilateral.

4.5.5. *Axisymmetric problems*

Axisymmetric problems consider structures and loads that are symmetric about an axis. This class of problems is similar to the class of planar problems, since axisymmetric problems can be described by planar models (see Figure 4.10) [DHA 81].

Figure 4.10. *Axisymmetric problem*

Consider a cylindrical coordinate system ρ, θ and z. The displacement in the θ-direction is zero, and every parameter is independent of θ by symmetry. In cylindrical coordinates, ∇u is defined by the formula:

$$\nabla u = \begin{pmatrix} \frac{\partial u}{\partial \rho} & \frac{1}{\rho}\left(\frac{\partial u}{\partial \theta} - v\right) & \frac{\partial u}{\partial z} \\ \frac{\partial v}{\partial \rho} & \frac{1}{\rho}\left(\frac{\partial v}{\partial \theta} + u\right) & \frac{\partial v}{\partial z} \\ \frac{\partial w}{\partial \rho} & \frac{1}{\rho}\frac{\partial w}{\partial \theta} & \frac{\partial w}{\partial z} \end{pmatrix}, \qquad [4.54]$$

where u, v, and w represent the displacement in the ρ, θ, and z-directions, respectively.

As $v = 0$ and $\frac{\partial u}{\partial \theta} = \frac{\partial w}{\partial \theta} = 0$,

$$\nabla^s u = \begin{pmatrix} \frac{\partial u}{\partial \rho} & 0 & \frac{1}{2}\left(\frac{\partial u}{\partial z} + \frac{\partial w}{\partial \rho}\right) \\ 0 & \frac{u}{\rho} & 0 \\ \frac{1}{2}\left(\frac{\partial u}{\partial z} + \frac{\partial w}{\partial \rho}\right) & 0 & \frac{\partial w}{\partial z} \end{pmatrix}. \qquad [4.55]$$

The strain tensor therefore has 4 components. The same is true for σ, since $\gamma_{\rho\theta} = \gamma_{z\theta} = 0$, which implies that $\sigma_{\rho\theta} = \sigma_{z\theta} = 0$. In vector notation:

$$\sigma = \begin{pmatrix} \sigma_{\rho\rho} \\ \sigma_{\theta\theta} \\ \sigma_{zz} \\ \sigma_{\rho z} \end{pmatrix} \qquad \epsilon = \begin{pmatrix} \epsilon_{\rho\rho} \\ \epsilon_{\theta\theta} \\ \epsilon_{zz} \\ \epsilon_{\rho z} \end{pmatrix} = \begin{pmatrix} \frac{\partial u}{\partial \rho} \\ \frac{u}{\rho} \\ \frac{\partial w}{\partial z} \\ \frac{\partial u}{\partial z} + \frac{\partial w}{\partial \rho} \end{pmatrix}. \qquad [4.56]$$

The matrix B has the following form:

$$\epsilon = \begin{pmatrix} < \frac{\partial N_i}{\partial \rho} > & < 0 > \\ < \frac{N_i}{\rho} > & < 0 > \\ < 0 > & < \frac{\partial N_i}{\partial z} > \\ < \frac{\partial N_i}{\partial z} > & < \frac{\partial N_i}{\partial \rho} > \end{pmatrix} \begin{pmatrix} u_i \\ w_i \end{pmatrix} = B u_n. \qquad [4.57]$$

From Hooke's law, we can find an elastic matrix that relates σ and ϵ. This matrix is of the following form (assuming isotropic elasticity):

$$A = \frac{E(1-\nu)}{(1+\nu)(1-2\nu)} \begin{pmatrix} 1 & \frac{\nu}{1-\nu} & \frac{\nu}{1-\nu} & 0 \\ \dots & 1 & \frac{\nu}{1-\nu} & 0 \\ \dots & \dots & 1 & 0 \\ \dots & \dots & \dots & \frac{1-2\nu}{2(1-\nu)} \end{pmatrix}. \qquad [4.58]$$

As the problem is axially symmetric, the elementary stiffness matrix is:

$$K^e = 2\pi \int_{\Omega^r} B^t(r,s) A B(r,s) \det(J) \rho(r,s) dr ds. \qquad [4.59]$$

We can use numerical integration to compute both this matrix and the stress matrix:

$$F^e = 2\pi \int_{\Omega^r} N^t(r,s)\ f\ \det(J)\ \rho(r,s) dr ds. \qquad [4.60]$$

Various types of elements can be used. For example, we can use the same elements for planar problems, with the interpolating functions remaining the same.

In general, the elements can be divided into three groups: planar stresses, planar strains and axisymmetrics. The stiffness matrix is computed differently in each case.

4.5.6. *Three-dimensional problems*

For three-dimensional problems, the stress and strain tensors are complete [CIA 88]:

$$\sigma = \begin{pmatrix} \sigma_{11} \\ \sigma_{22} \\ \sigma_{33} \\ \sigma_{12} \\ \sigma_{13} \\ \sigma_{23} \end{pmatrix} \quad \epsilon = \begin{pmatrix} \epsilon_{11} \\ \epsilon_{22} \\ \epsilon_{33} \\ \gamma_{12} \\ \gamma_{13} \\ \gamma_{23} \end{pmatrix}. \qquad [4.61]$$

We can use either the isotropic or the anisotropic version of Hooke's law to determine the behavior matrix. The elements are usually chosen from one of the three shapes: tetrahedra, hexahedra and prisms.

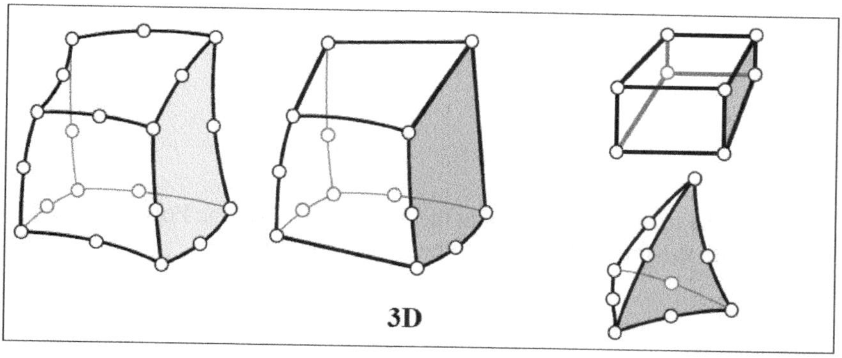

Figure 4.11. *3D elements*

4.6. Using *Matlab*

The latest version of *Matlab* features a PDE Toolbox, which can be accessed by running the *pdftool* command. This opens the window shown in [4.12], which allows the data of the problem to be entered [MAT 14].

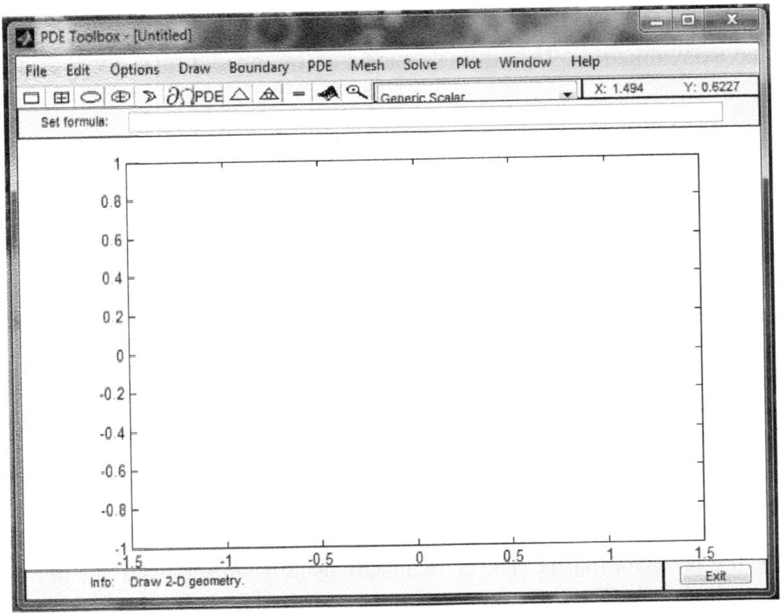

Figure 4.12. *Window of the PDE Toolbox*

4.6.1. *Solving Poisson's equation*

We can use the *assempde* function in the Toolbox to solve the Poisson problem on the unit disk with the following Dirichlet boundary conditions:

$$\begin{cases} -\Delta u = 1 & \text{on } \Omega \\ u = 0 & \text{on } \partial\Omega \end{cases} \qquad [4.62]$$

The script listed below can be used to solve this problem numerically:

```
g='circleg'; b='circleb1'; c=1; a=0; f=1;
[p,e,t]=initmesh(g,'hmax',1); figure; pdemesh(p,e,t);

axis equal er = Inf; while er > 0.001
    [p,e,t]=refinemesh(g,p,e,t);
    u=assempde(b,p,e,t,c,a,f);
    exact=(1-p(1,:).^2-p(2,:).^2)'/4;
    er=norm(u-exact,'inf');
    fprintf('Error: %e. Number of nodes: %d\n',er,size(p,2));
```

```
end figure; pdemesh(p,e,t); axis equal figure; pdesurf(p,t,u-exact);
figure; pdesurf(p,t,u);
```

Figures 4.13–4.16 show the results of this computation.

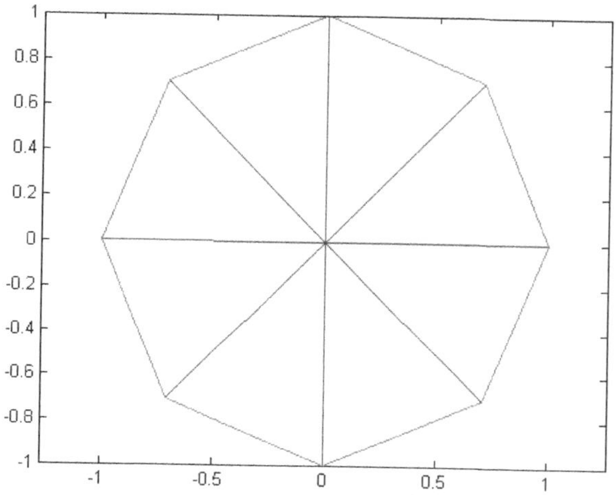

Figure 4.13. *Discretization of the domain. For a color version of this figure, see www.iste.co.uk/radi/advanced2.zip*

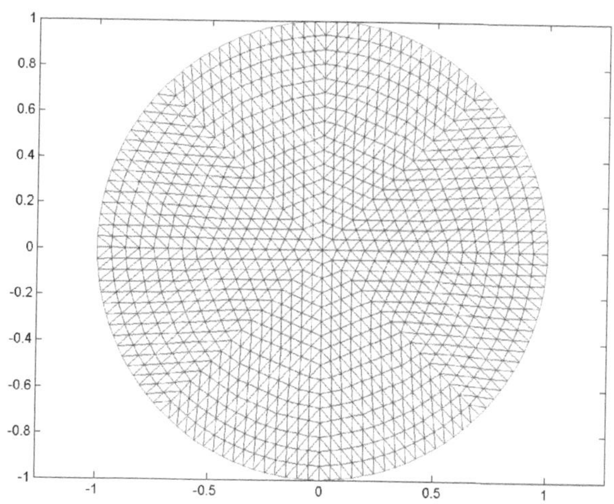

Figure 4.14. *Mesh of the domain. For a color version of this figure, see www.iste.co.uk/radi/advanced2.zip*

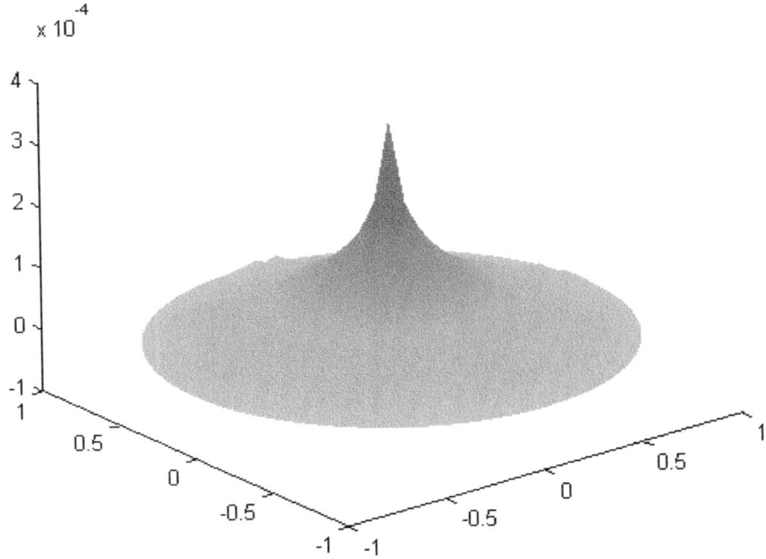

Figure 4.15. *Computed solution. For a color version of this figure, see www.iste.co.uk/radi/advanced2.zip*

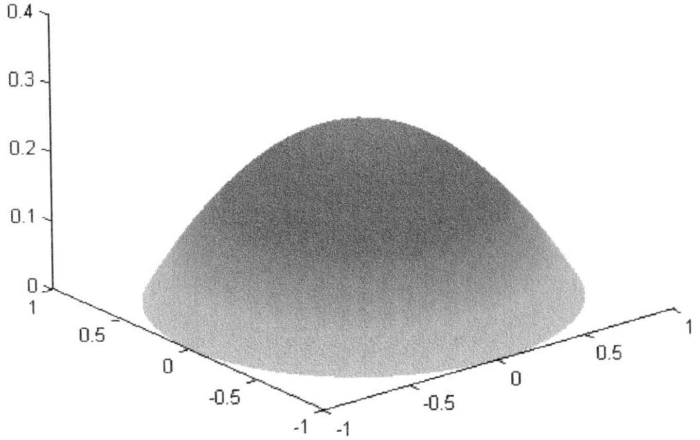

Figure 4.16. *Comparison of the computed solution and the exact solution. For a color version of this figure, see www.iste.co.uk/radi/advanced2.zip*

4.6.2. *Solving the heat equation*

Suppose that we wish to solve the heat equation in the following problem [KOK 09]:

Find u: $[0,1] \times [0,T] \to \mathbb{R}$, such that:

$$\frac{\partial u}{\partial t}(x,t) - \frac{1}{\pi^2}\frac{\partial^2 u}{\partial x^2}(x,t) = 0,\ x \in]0,1[,\ t \in]0,T[\qquad [4.63]$$

$$u(0,t) = u(1,t) = 0, \qquad t \in [0,1] \qquad [4.64]$$

$$u(x,0) = \cos\left(\pi\left(x - \frac{1}{2}\right)\right), \qquad x \in]0,1[. \qquad [4.65]$$

The exact solution of Problem [4.63–4.65] is $u(x,t) = e^{-t}\cos(\pi(x-\frac{1}{2}))$ for all $(x,t) \in [0,1] \times [0,T]$. To compute an approximate solution for this equation, we can discretize in time using a θ-scheme, then discretize in space with finite elements.

We can use the following *Matlab* script to do this:

```
%--------------------------------------------------
%Solving the heat equation
%--------------------------------------------------
%
%subdivisions space /time
ht=0.01; Tmax=1.2; nx=33; hx=1/(nx-1); x=[0:hx:1]';

%matrices
K=stiff(1/pi^2,hx,nx); M=mass(1/ht,hx,nx); A=M+K;

%boundary conditions by deleting rows
A(nx,:)=[]; A(:,nx)=[]; A(1,:)=[]; A(:,1)=[];

%creation
R=chol(A);

%initial condition
u=cos(pi*(x-1/2));

%time step loop
k=0; while (k*ht<Tmax)
    k=k+1;
    %compute the right-hand side
```

```
    b=M*u;
    b(nx)=[];b(1)=[];
    %solve
    u=zeros(nx,1);
    u(2:nx-1)=R\(R'\b);
end
```

This script calls the two subfunctions *stiff* and *mass*, which can, for example, be implemented as follows:

```
function R=stiff(nu,h,n)
%
[m1,m2]=size(nu);

if (length(nu)==1)
    ee=nu*ones(n-1,1)/h;  e1=[ee;0]; e2=[0;ee]; e=e1+e2;
else
    ee=.5*(nu(1:n-1)+nu(2:n))/h; e1=[ee;0]; e2=[0;ee]; e=e1+e2;
end

R=spdiags([-e1 e -e2],-1:1,n,n);
return

function M=mass(alpha,h,n)

[m1,m2]=size(alpha);

if(length(alpha)==1)
    ee=alpha*h*ones(n-1,1)/6;
    e1=[ee;0]; e2=[0;ee]; e=e1+e2;
else
    ee=h*(alpha(1:n-1)+alpha(2:n))/12;
    e1=[ee;0]; e2=[0;ee]; e=e1+e2;
end

M=spdiags([e1 2*e e2],-1:1,n,n);
return
```

4.6.3. *Computing structures*

Consider the structure shown in Figure 4.17, with the following parameters $E = 200\,000$ MPa, cross-section $A = 6\,000$ mm^2, and moment of inertia $I = 200 \times 10^6$ mm^4.

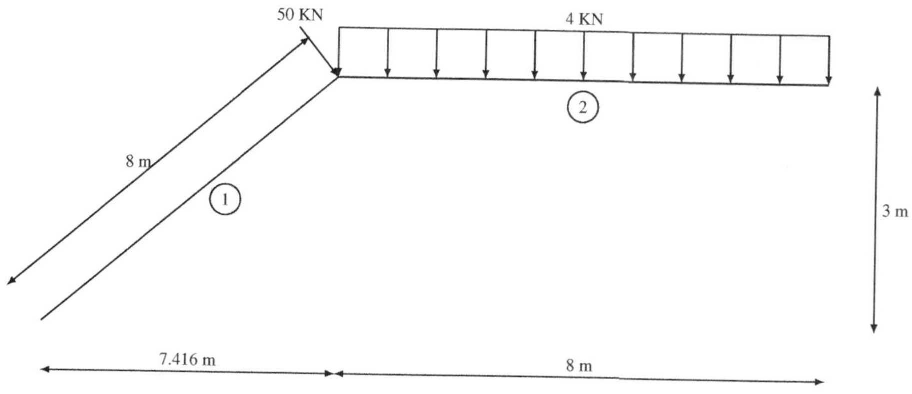

Figure 4.17. *Structure of the problem*

The following *Matlab* script computes the structure of this problem using the finite element method:

```
% Set the matrices to zero
k=zeros(6,6,2); K=zeros(6,6,2); Gamma=zeros(6,6,2);
% Enter parameter values, in units: mm^2, mm^4, and MPa(10^6 N/m)
A=6000; II=200*10^6; EE=200000;
% Convert units into meters and kN
A=A/10^6; II = II/10^12; EE =EE*1000;
% Element 1
i=[0,0]; j=[7.416,3];
[k(:,:,1),K(:,:,1),Gamma(:,:,1)]=stiff(EE,II,A,i,j);
% Element 2
i=j; j=[15.416,3];
[k(:,:,2),K(:,:,2),Gamma(:,:,2)]=stiff(EE,II,A,i,j);
% Define the elementary stiffness matrix
ID=[-4 1 -7;-5 2 -8;-6 3 -9];
% Define the connection matrix
LM=[-4 -5 -6 1 2 3; 1 2 3 -7 -8 -9];
% Assemble the augmented stiffness matrix
Kaug = zeros(9); for elem=1:2
    for r=1:6
        lr=abs(LM(elem,r));
        for c=1:6
            lc=abs(LM(elem,c));
            Kaug(lr,lc)=Kaug(lr,lc)+K(r,c,elem);
        end
    end
```

```
end
% Extract the stiffness matrix
Ktt=Kaug(1:3,1:3);

% Determine the reactions at the nodes in local coordinates
fea(1:6,1)=0; fea(1:6,2)=[0,8*4/2,4*8^2/12,0,8*4/2,-4*8^2/12]';
% Determine the reactions in the global coordinate system
FEA(1:6,1)=Gamma(:,:,1)'*fea(1:6,1);
FEA(1:6,2)=Gamma(:,:,2)'*fea(1:6,2);
% FEA_Rest for constrained nodes
FEA_Rest=[0,0,0,FEA(4:6,2)'];
% Assemble the right-hand side for non-constrained nodes
P(1)=50*3/8; P(2)=-50*7.416/8-FEA(2,2); P(3)=-FEA(3,2);
% Solve to find the displacements in meters and in radians
Displacements=inv(Ktt)*P'
% Extract Kut
Kut = Kaug(4:9,1:3);
% Compute the reactions and introduce boundary conditions
Reactions=Kut*Displacements+FEA_Rest'
% Solve to find the internal forces, excluding fixed points
dis_global(:,:,1)=[0,0,0,Displacements(1:3)'];
dis_global(:,:,2)=[Displacements(1:3)',0,0,0]; for elem=1:2
    dis_local= Gamma(:,:,elem)*dis_global(:,:,elem)';
    int_forces= k(:,:,elem)*dis_local+fea(1:6,elem)
end
```

The above script calls the *stiff* function, which can be implemented as follows:

```
function [k,K,Gamma] = stiff( EE,II,A,i,j )
% Find the length
L=sqrt((j(2)-i(2))^2+(j(1)-i(1))^2);
% Compute the angle theta
if(j(1)-i(1))~=2
    alpha=atan((j(2)-i(2))/(j(1)-i(1)))
else
    alpha=-pi/2;
end
% Form the rotation matrix Gamma
Gamma =[cos(alpha) sin(alpha) 0 0 0 0;

-sin(alpha) cos(alpha) 0 0 0 0;
```

```
    0 0 1 0 0 0;
    0 0 0 cos(alpha) sin(alpha) 0;
    0 0 0 -sin(alpha) cos(alpha) 0;
    0 0 0 0 0 1];
% Form the elementary stiffness matrix in local coordinates
EI=EE*II; EA=EE*A; k=[EA/L, 0, 0, -EA/L, 0, 0; 0, 12*EI/L^3,
6*EI/L^2, 0, -12*EI/L^3,6*EI/L^2;
    0, 6*EI/L^2, 4*EI/L, 0 -6*EI/L^2, 2*EI/L;
    -EA/L, 0 ,0 , EA/L, 0, 0;
    0, -12*EI/L^3, -6*EI/L^2, 0, 12*EI/L^3, -6*EI/L^2;
    0, 6*EI/L^2, 2*EI/L, 0, -6*EI/L^2, 4*EI/L];
% Elementary matrix in global coordinates
K=Gamma'*k*Gamma;
end
```

Executing this script returns the following results:

```
>> finite_elements

alpha =

    0.3844

alpha =

     0

Displacements =

    0.0010
   -0.0050
   -0.0005

Reactions =

  130.4973
   55.6766
   13.3742
 -149.2473
   22.6734
  -45.3557
```

```
int_forces =

  149.2473
    9.3266
   -8.0315
 -149.2473
   22.6734
  -45.3557
```

5

Finite Volume Methods

5.1. Introduction

Finite volume methods are a class of numerical analysis methods used to solve PDEs numerically, much like the finite element method and finite difference methods. However, unlike finite difference methods, which approximate the derivatives, finite volume methods and the finite element method approximate the integrals and then apply Gauss's divergence theorem. Finite volume methods work directly from the so-called strong form of the equation, whereas finite element methods are based on a variational formulation.

The idea of the technique of control volumes is to integrate the PDE on a certain set of control volumes to obtain discretized equations that conserve the value of every physical quantity on each volume.

Every conservation equation is of the same form, and can be stated in terms of a single general formula for the transport equation of a scalar property. In the notation popularized by Bird, Steward and Lightfoot [BIR 06] and later reused by Brodkey and Hershey, two of the best-known researchers in transport phenomena, this formula can be presented as:

$$\frac{\partial \rho\phi}{\partial t} + \nabla \cdot J = S^{\phi}, \qquad [5.1]$$

where ϕ denotes a scalar representing some property; J summarizes the convection and diffusion flux terms of ϕ, defined as $J = \rho\phi u \Gamma_{\phi} \nabla\phi$, with Γ_{ϕ} the diffusion coefficient of the variable ϕ and S^{ϕ} is the source term of ϕ.

This convection-diffusion equation can also be written in the following form:

$$\frac{\partial \rho\phi}{\partial t} + \mathrm{div}(\rho\phi u) = \mathrm{div}(\Gamma_{\phi}\nabla\phi) + S^{\phi}. \qquad [5.2]$$

It intrinsically contains multiple transport equations in the dependent variable ϕ. In general, each term has a specific, well-defined physical interpretation [LEV 02].

For example, if $\phi = 1$ and $S^\phi = 0$, then we recover the continuity equation:

$$\frac{\partial \rho}{\partial t} + \mathrm{div}(\rho u) = 0. \tag{5.3}$$

If $\phi = T$, $S^\phi = Q(T)$ and $\Gamma_\phi = k/c_p$, then we recover a simplified version of the energy equation:

$$\frac{\partial \rho T}{\partial t} + \mathrm{div}(\rho u T) = \mathrm{div}\left(\frac{K}{c_p}\nabla T\right) + Q(T). \tag{5.4}$$

Most equations describing the transport of a quantity can be derived similarly. The generalized transport equation is written in the form of a divergence, which allows us to apply Gauss's theorem when working with the integral equations.

5.2. Finite volume method (FVM)

The finite volume method (FVM) can be viewed as a special case combining both the finite difference method and the method of mean weighted residuals (variables of the FEM). The underlying idea of the FVM is easy to understand and lends itself well to direct physical interpretation: the computation domain is divided into distinct (disjoint) elements, called "control volumes" (CVs), in such a way that each control volume encloses one node of the computation mesh. The PDE is then integrated on each control volume. To compute the integral, the variable is approximated using profile functions (e.g. piecewise linear or quadratic) between the points of the computation mesh. This leads to a discretization of the PDE whose unknowns are the values of the variables on the set of mesh points.

5.2.1. *Conservation properties of the method*

One of the most important aspects of the FVM is its conservation properties, which rely heavily on the integral formulation of the method. Specifically, in the integral form of the general transport equation:

$$\frac{\partial}{\partial t}\int_\Omega \rho\phi d\Omega + \oint_\Gamma J \cdot n d\Gamma = \int_\Omega S^\phi d\Omega, \tag{5.5}$$

the variation in the property ϕ is a function of the net flux that passes through the surface Γ enclosing the volume Ω. This property is crucial: after partitioning the domain into multiple subdomains with arbitrarily many sides, the following result holds:

$$\int_A^B J \cdot n d\Gamma = -\int_B^A J \cdot n d\Gamma, \qquad [5.6]$$

where A and B are the two end points of each side.

In other words, regardless of the choice of partition, the internal flux terms always cancel, and the value of the global property ϕ is conserved.

5.2.2. *The stages of the method*

To solve the generalized transport equation by the FVM, we apply the volume and contour integrals to a certain set of elements and then approximate the integrals with algebraic expressions. Thus, the first step is to divide the domain into a finite set of subdomains. Then, we compute the value of the variable ϕ on each element, or control volume. The individual steps of the method can be summarized as follows:

– partition the region into subdomains or control volumes. These control volumes must cover the domain completely and can have various shapes. The most common basis elements are quadrilaterals and triangles in two dimensions, and hexahedra and tetrahedra in three dimensions;

– integrate the equations on each control volume and apply Gauss's divergence theorem. On each control volume, we replace the function ϕ by an approximation; this is typically a constant fixed to the center of gravity of the element. We also estimate the flux through the sides of the control volume, usually in the form of an average value;

– incorporate the boundary conditions;

– compute the balance of ϕ on each volume;

– solve the algebraic system thus obtained.

In this chapter, we will demonstrate the FVM by applying it to examples of fluid flow problems, beginning with a scalar transport equation and then moving on to the Navier–Stokes equations.

We recall that the numerical methods used to solve a PDE in the context of a given model must satisfy certain conditions, such as convergence, consistency and stability.

5.2.3. *Convergence*

An approximation scheme is said to be convergent if the numerical solution approaches the continuous problem (the solution of the PDE) as the space and time steps tend to zero. In other words, a scheme is convergent if the errors resulting from the spatiotemporal discretization decrease with the spatiotemporal step size while remaining bounded.

In practice, it is not easy to verify that this property is satisfied. Instead, we consider methods for which we can study two fundamental properties, consistency and stability, and apply the Lax equivalence theorem, which states that if a PDE is approximated by a consistent linear scheme, then this scheme converges if and only if it is stable.

5.2.4. *Consistency*

A finite difference scheme is said to be consistent with the original PDE if the truncation error tends to zero as the discretization step ($\Delta x, \Delta y, \Delta z$ for space, and Δt for time) tends to zero (in other words, as the discretization steps approach zero, the scheme describing the discrete problem transforms into the equation that describes the continuous problem). It is easy to study whether a given scheme satisfies this property by considering Taylor series.

5.2.5. *Stability*

The objective of rewriting a PDE in terms of finite differences is to allow us to solve it numerically, but the solution that we obtain is of course only a numerical approximation. Any such approximation is only satisfactory if it tends to the exact solution as the integration step tends to zero. This fundamental convergence condition leads us to define the condition of stability – a scheme is stable if the solution is bounded whenever the initial condition is bounded, for sufficiently small Δt and ($n\Delta t < \infty$).

It can be tricky to show that a given numerical scheme is stable, especially for nonlinear schemes. Several methods for studying stability have been suggested, but a general approach is not available; the most commonly adopted strategy is von Neumann analysis, which does not consider the boundary conditions of the PDE when studying whether the stability condition is satisfied.

To solve the problem of numerical stability, an alternative approach has been suggested by several authors: the idea is to tackle the stability problem (whether or not the scheme is stable) early in the design stage by imposing certain properties, such as conservation properties, positivity or the TVD (total variation decreasing) property. The numerical scheme must satisfy the following conditions:

– the transported quantity must be conserved;

– if the transported quantity increases at a given point by convection and diffusion, then it must not decrease at neighboring points;

– the diffusion must remain linear.

5.3. Advection schemes

Below, we present a numerical method based on this integral formulation (method of finite volumes/control volumes). As an example, we will study the simplest possible transport equation, the unsteady pure diffusion transport equation:

$$\mathrm{div}(\Gamma \vec{grad}\phi) + S_\phi = 0. \qquad [5.7]$$

Formally integrating this equation (the key stage of the FVM) over a control volume and then applying the divergence theorem yields:

$$\int_{VC} \mathrm{div}(\Gamma \vec{grad}\phi) dv + \int_{VC} S_\phi dv = \int_A \vec{n}(\Gamma \vec{grad}\phi) dA + \int_{VC} S_\phi dv = 0.[5.8]$$

To illustrate more clearly how the FVM works, we will consider the one-dimensional steady-state diffusion equation on $[A, B]$:

$$\frac{d}{dx}\left(\Gamma \frac{d\phi}{dx}\right), \qquad [5.9]$$

where Γ is the diffusion coefficient, S_ϕ is a source term and the values of θ at the end points A and B are given, respectively, by $\phi_A = \phi(A)$ and $\phi_B = \phi(B)$:

Step 1. Meshing

The first step is to generate the mesh (computation grid). This involves dividing the computational domain into finitely many control volumes (elements, cells, etc.). To implement the finite volume method, we distribute a set of points, the nodes, between the end points A and B of the domain. The boundaries of each control volume are positioned between pairs of adjacent nodes; in other words, each node is enclosed by one control volume or cell. It is often useful to choose the boundaries of the control volume at the edges of the domain (here, A and B) to coincide with the boundaries.

Figure 5.2 shows the positioning of the nodes, the control volumes, their boundaries and that each control volume has “length” Δx.

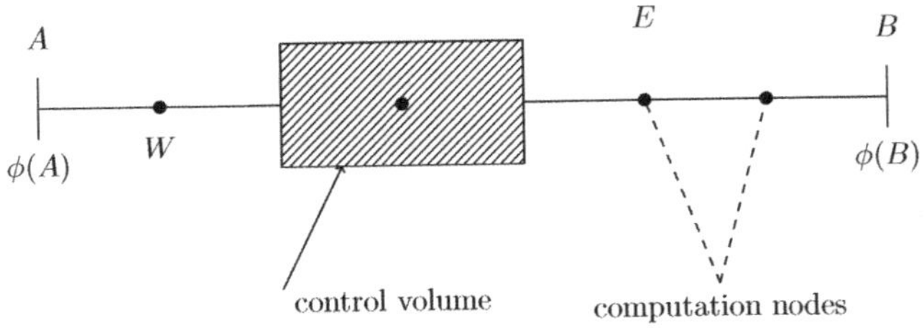

Figure 5.1. *Control volume*

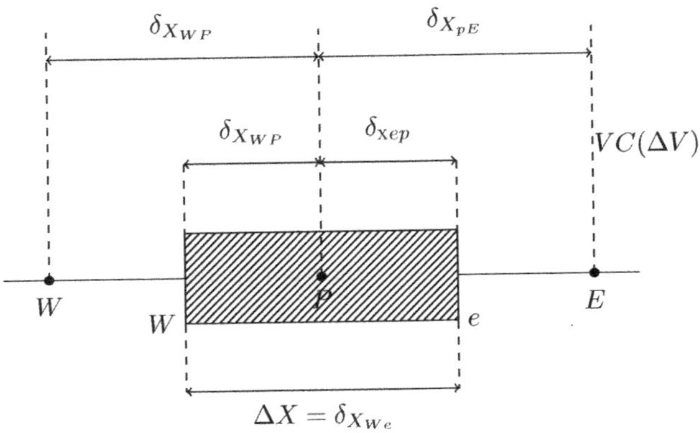

Figure 5.2. *Positioning of nodes and control volumes*

Step 2. Discretization

As mentioned earlier, the most important stage of the FVM is integrating the PDE on each control volume to obtain discretized equations at the node P. For the control volume Δv defined in the figure:

$$\int_{VC} \frac{d}{dx}(\Gamma \frac{d\phi}{dx})dv + \int_{VC} S_\phi dv = (\Gamma A \frac{d\phi}{dx})_e - (\Gamma A \frac{d\phi}{dx})_w + \bar{S}\Delta v = 0, \quad [5.10]$$

where A is the surface whose sides are perpendicular to $\bar{ox}$, Δv is the control volume and $\bar{S}$ is the average value of the source term S on the control volume.

This discretized equation has a clear physical interpretation; the equation states that the diffusive flux of ϕ exiting the control volume via the "East e" face minus the diffusive flux of ϕ entering the control volume via the "West w" face is equal to the quantity of ϕ generated by the source term inside the control volume. In other words, this equation describes the balance in the quantity of ϕ over the control volume.

To complete the discrete equation, we still need to evaluate the diffusion coefficient r and the gradient $\frac{d\phi}{dx}$ at the points "e" and "w", which are not mesh nodes (fictitious points). We only know the values of Γ and $\frac{d\phi}{dx}$ at the nodes W, P and E, so we need to approximate these variables using their values at these nodes. For example, the diffusion coefficient at the "w, e" interfaces of the control volume can be calculated by linear interpolation:

$$\Gamma_e = \frac{\Gamma_E + \Gamma_P}{2}; \Gamma_w = \frac{\Gamma_W + \Gamma_P}{2}. \qquad [5.11]$$

The diffusive flux term can be calculated by:

$$\left(\Gamma A \frac{d\phi}{dx}\right)_w = \Gamma_w A_w \frac{\phi_P - \phi_W}{\delta x_{pw}} \quad ; \quad \left(\Gamma A \frac{d\phi}{dx}\right)_e = \Gamma_e A_e \frac{\phi_E - \phi_P}{\delta x_{pE}}. \qquad [5.12]$$

In practice, the source term is usually a function of the variable ϕ. If so, the FVM approximates the source term by a linear relation:

$$\bar{S}\Delta v = S_u + S_p \phi_p. \qquad [5.13]$$

Finally, we have that:

$$\Gamma_w A_w \frac{\phi_P - \phi_W}{\delta x_{pw}} + \Gamma_e A_e \frac{\phi_E - \phi_P}{\delta x_{pE}} \qquad [5.14]$$

$$\underbrace{\left(\frac{\Gamma_e A_e}{\delta x_{pE}} + \frac{\Gamma_w A_w}{\delta x_{pw}} - S_p\right) \phi_p}_{a_p \phi_p} = \underbrace{\frac{\Gamma_w A_w}{\delta x_{pw}} \phi_w}_{a_w \phi_w} + \underbrace{\frac{\Gamma_e A_e}{\delta x_{pE}} \phi_E}_{a_E \phi_E} + S_u \qquad [5.15]$$

The following relation holds:

$$a_p = a_E + a_w - S_p. \qquad [5.16]$$

The terms S_u, S_p are known and can be calculated from a model that approximates the source term S.

REMARK.– The expressions of ϕ at the interfaces w and e are indirectly assumed to have a piecewise linear profile in these computations.

Step 3. Solving the system

The discrete equation obtained above for a single point P must be satisfied by every point on the computational mesh. For the control volumes at the physical boundaries of the domain, this equation needs to be modified to account for the boundary conditions. The resulting algebraic system is then solved by a suitable choice of algorithm to obtain a solution on the nodes of the mesh.

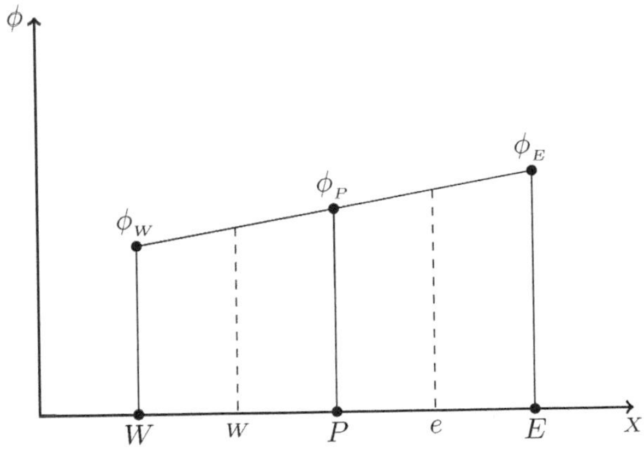

Figure 5.3. *The function* ϕ

EXAMPLE.– Consider a bar AB of length $L = 50\,cm$ and cross-section $A = 10^{-2}m^2$. Suppose that we wish to compute the temperature distribution along the bar, assuming a thermal conductivity of $k = 1,000W/m/k$ and subject to the conditions that the temperature is $T_A = T(A) = 100$ at A and $T_B = T(B) = 500$ at B. The heat conduction equation is $\frac{d}{dx}\left(k\frac{dT}{dx}\right) = 0$.

We will consider five nodes and a uniform mesh $\delta x = 0.1\,m$.

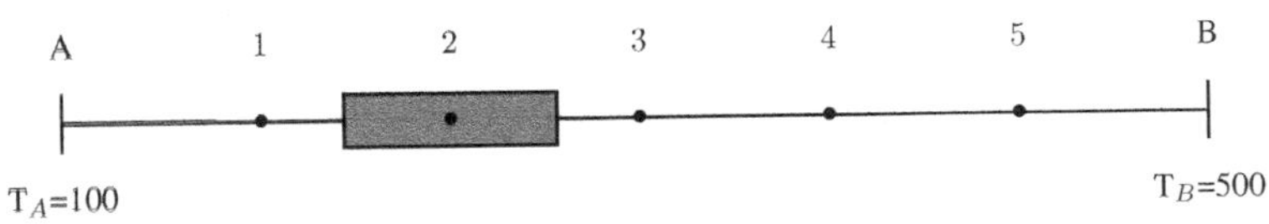

Figure 5.4. *Discretization of the function* T

– For each of the interior nodes $P = 2, 3, 4$, as $k_e = K_w = k$ and $A_e = A_w = A$:

$$\begin{cases} a_p T_p = a_W T_W + a_E T_E \\ a_E = \dfrac{kA}{\delta x} = a_W \\ a_p = a_E + a_W \end{cases} \qquad [5.17]$$

– The nodes 1 and 5 at the boundary need to be handled differently to account for the boundary conditions.

NODE 1 (P=1).–

$$K_A \frac{T_E - T_p}{\delta x} - k_A \frac{T_p - T_A}{\delta x/2} = 0 \qquad [5.18]$$

The term $\delta x/2$ comes from the fact that the distance from the boundary A to the node $P = 1$ is equal to half of the mesh size. In other words:

$$\left(\frac{KA}{\delta x} + \frac{2KA}{\delta x}\right) T_p = 0.T_W + \frac{KA}{\delta x}\left(\frac{KA}{\delta x}\right) T_E + \left(\frac{2KA}{\delta x}\right) T_A, \qquad [5.19]$$

where $a_w = 0$, $a_E = \dfrac{kA}{\delta x}$, $S_p = -\dfrac{2kA}{\delta x}$, $S_u = \dfrac{2KA}{\delta x} T_A$ and $a_p = a_E + a_W - S_p$.

We do the same for node 5 at the other boundary:

NODE 5 ($P = 5$).–

$$k_A \frac{T_B - T_p}{\delta x/2} - k_A \frac{T_p - T_W}{\delta x} = 0 \qquad [5.20]$$

$$\left(\frac{k_A}{\delta x} + \frac{2k_A}{\delta x}\right) T_p = \frac{kA}{\delta x} T_W + 0.T_E + \frac{2kA}{\delta x} T_B, \qquad [5.21]$$

where $a_w = \dfrac{kA}{\delta x}$, $a_E = 0$, $S_p = -\dfrac{2kA}{\delta x}$, $S_u = \dfrac{2KA}{\delta x} T_A$ and $a_p = a_E + a_W - S_p$.

The resulting linear system can now be written as follows:

$$\frac{k_A}{\delta x} = 100$$

– Node 1: $300T_1 = 100T_2 + 200T_A$

– Node 2: $20073 = 100T_l + 100T_3$

– Node 3: $200T3 = 100T_2 + 100T_4$

– Node 4: $200T4 = 100T_3 + 200T_5$

– Node 5: $300T5 = 100T_4 + 200T_B$

In matrix form:

$$\begin{bmatrix} 300 & -100 & 0 & 0 & 0 \\ -100 & 200 & -100 & 0 & 0 \\ 0 & -100 & 200 & -100 & 0 \\ 0 & 0 & -100 & 200 & -100 \\ 0 & 0 & 0 & -100 & 200 \end{bmatrix} \begin{Bmatrix} T_1 \\ T_2 \\ T_3 \\ T_4 \\ T_5 \end{Bmatrix} = \begin{Bmatrix} 200T_A \\ 0 \\ 0 \\ 0 \\ 200T_B \end{Bmatrix}. \qquad [5.22]$$

This method can be solved with "any old" numerical method for solving linear systems. The solution is:

$$\begin{cases} T_1 = 140 \\ T_2 = 220 \\ T_3 = 300 \\ T_4 = 380 \\ T_5 = 480 \end{cases} \qquad [5.23]$$

By comparison, the exact solution is $T(x) = 800x + 100$.

5.3.1. *Two-dimensional FVM*

To illustrate the FVM in two dimensions, we will consider the two-dimensional steady-state diffusion equation:

$$\frac{\partial}{\partial x}\left(\Gamma\frac{\partial \phi}{\partial x}\right) + \frac{\partial}{\partial y}\left(\Gamma\frac{\partial \phi}{\partial y}\right) + S = 0. \qquad [5.24]$$

For the two-dimensional problem, as well as the nodes E, W, we consider the two additional nodes N, S in the direction of $\vec{oy}$.

Integrating the diffusion equation over a control volume gives Δv:

$$\int_{\Delta v} \frac{\partial}{\partial x}\left(\Gamma\frac{\partial \phi}{\partial x}\right) dxdy + \int_{\Delta v} \frac{\partial}{\partial y}\left(\Gamma\frac{\partial \phi}{\partial y}\right) dxdy + \int_{\Delta v} S_\phi dv = 0. \qquad [5.25]$$

We note that, here, $A_e = A_w = \Delta y$; $A_n = A_s = \Delta x$, which gives:

$$\left[\Gamma_e A_e \left(\frac{\partial \phi}{\partial x}\right)_e - \Gamma_w A_w \left(\frac{\partial \phi}{\partial x}\right)_w\right] + \left[\Gamma_n A_n \left(\frac{\partial \phi}{\partial y}\right)_n - \Gamma_s A_s \left(\frac{\partial \phi}{\partial y}\right)_s\right] + \bar{S}\Delta v = 0. \quad [5.26]$$

As in the one-dimensional case, this equation can be interpreted as the conservation of ϕ within the control volume, describing the balance of the quantity of ϕ generated (by the source) and the various flux terms through each interface of Δv. With similar approximations as in the one-dimensional case, we obtain:

– flux through the "West w" face:

$$\Gamma_w A_w \left(\frac{\partial \phi}{\partial x}\right)_w = \Gamma_w A_w \frac{\phi_P - \phi_W}{\delta x_{pW}} \quad [5.27]$$

– flux through the "East e" face:

$$\Gamma_e A_e \left(\frac{\partial \phi}{\partial x}\right)_e = \Gamma_e A_e \frac{\phi_E - \phi_P}{\delta x_{pE}} \quad [5.28]$$

– flux through the "North n" face:

$$\Gamma_n A_n \left(\frac{\partial \phi}{\partial y}\right)_n = \Gamma_n A_n \frac{\phi_N - \phi_P}{\delta x_{Np}} \quad [5.29]$$

– flux through the "South s" face:

$$\Gamma_s A_s \left(\frac{\partial \phi}{\partial y}\right)_s = \Gamma_s A_s \frac{\phi_P - \phi_S}{\delta x_{sp}} \quad [5.30]$$

and also:

$$\Gamma_w = \frac{\Gamma_P + \Gamma_w}{2}; \quad \Gamma_e = \frac{\Gamma_E + \Gamma_P}{2}$$

$$\Gamma_s = \frac{\Gamma_P + \Gamma_s}{2}; \quad \Gamma_n = \frac{\Gamma_P + \Gamma_N}{2}$$

$$\bar{S} = S_u + S_p \phi_P$$

Hence, the discretized equation is:

$$a_P \phi_P = a_E \phi_E + a_W \phi_W + a_S \phi_S + a_N \phi_N + S_n \quad [5.31]$$

$$a_W = \frac{\Gamma_w A_w}{\delta x_{Wp}};\quad a_E = \frac{\Gamma_e A_e}{\delta x_{pE}};\quad a_N = \frac{\Gamma_n A_n}{\delta x_{pN}};\quad a_S = \frac{\Gamma_s A_s}{\delta x_{pS}}$$

$$a_p = a_E + a_w + a_s + a_N - S_p$$

In three dimensions, the discretized equation has a similar structure, with additional terms a_T and a_B.

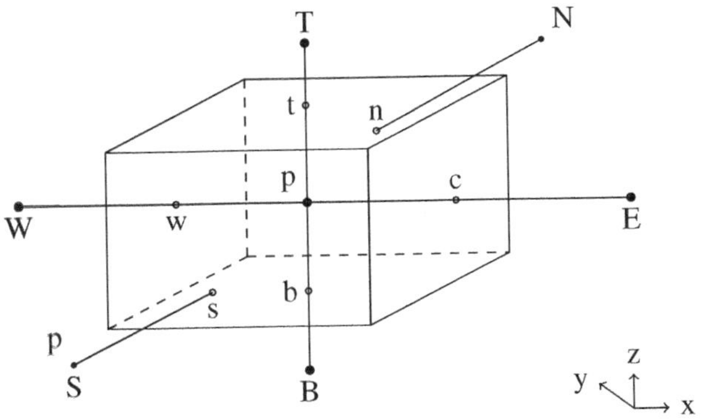

Figure 5.5. *Three-dimensional case*

In summary, for the steady-state diffusion equation:

– The discrete equation in one, two and three dimensions is of the following form in general:

$$a_p \phi_p = \sum a_{nb} \phi_{nb} + S_n, \qquad [5.32]$$

where the sum ranges over the nodes adjacent to P and a_{nb} denotes the corresponding coefficients $(a_w, a_E, a_N, a_s, a_B, a_T)$, ϕ_{nb} is the value of ϕ at the adjacent nodes and $(S_u + S_p\phi_p)$ is the linearized source term.

– In each case, the coefficients around the node P satisfy the relation:

$$a_p = \sum a_{nb} - S_p. \qquad [5.33]$$

– The source term is included in a linearized form:

$$\bar{S}\Delta v = S_u + S_p \phi_p. \qquad [5.34]$$

– The boundary conditions can either be incorporated exactly if the flux is imposed by $S_u + \phi_p S_p = q_B$, or as a linearized approximation by using additional coefficients S_u and S_p.

5.3.2. *Convection-diffusion equation*

Most fluid motion phenomena unfold according to two modes of transport: convection and diffusion [PAT 80]. In this section, we apply the FVM to a convection-diffusion transport equation (CDTE):

$$\mathrm{div}\left(\rho\vec{u}\phi\right) = \mathrm{div}\left(\Gamma \mathrm{grad}\phi\right) + S_\phi \qquad [5.35]$$

expressed in integral form over a control volume Δ_v:

$$\int_A \vec{n}\left(\rho\phi\vec{u}\right) dA = \int_A \vec{n}\left(\Gamma \mathrm{grad}\phi\right) dA + \int_{VC} S_\phi dv. \qquad [5.36]$$

The term on the left-hand side represents the convective flux, and the first term on the right-hand side represents the diffusive flux.

The primary difficulty associated with discretizing the convective term lies in choosing how to compute the value of the transported quantity at the interfaces of the control volume. The problem is that the convection process only exerts an influence in the direction of the flow, whereas the diffusion process acts in every direction through its gradients.

5.3.2.1. *One-dimensional steady-state CD equation*

Without a source term, the steady-state CD equation can be stated as follows for the quantity ϕ in one dimension:

$$\frac{d}{dx}\left(\rho u\phi\right) = \frac{d}{dx}\left(\Gamma\frac{d\phi}{dx}\right). \qquad [5.37]$$

The flow must satisfy the continuity equation:

$$\frac{d}{dx}\left(\rho u\right) = 0. \qquad [5.38]$$

Integrating the CD equation over the control volume (CV) gives the following expression (see Figure 5.6):

$$\left(\rho uA\phi\right)_e - \left(\rho uA\phi\right)_w = \left(\Gamma A\frac{\partial\phi}{\partial x}\right)_e - \left(\Gamma A\frac{\partial\phi}{\partial x}\right)_w. \qquad [5.39]$$

Integrating the continuity equation over the control volume gives:

$$\left(\rho AU\right)_e - \left(\rho AU\right)_w = 0. \qquad [5.40]$$

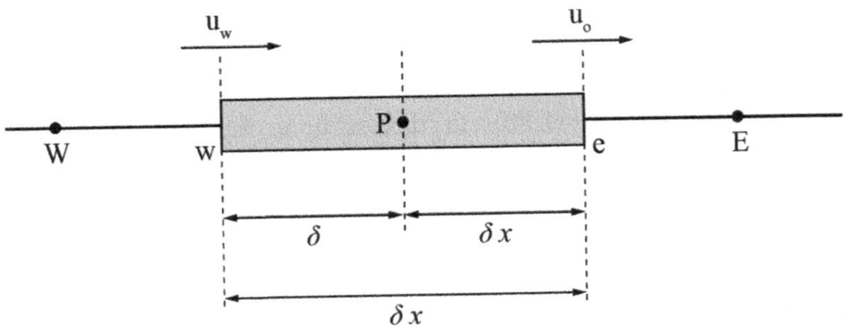

Figure 5.6. *Control volume*

To obtain the discrete equation, we still need to approximate each term at the points w and e, which are not on the computational mesh. To do this, we define two auxiliary variables:

$$F = \rho u \quad ; \quad D = \frac{\Gamma}{\delta x}. \tag{5.41}$$

Thus:

$$F_w = (\rho u)_w ; \quad D_w = \frac{\Gamma_w}{\delta x_{Wp}}$$

$$F_e = (\rho u)_e ; \quad D_e = \frac{\Gamma_e}{\delta x_{pE}}$$

With this notation, setting $A_e = A_w = A$, the discrete equation becomes:

$$F_e \phi_e - F_w \phi_w = D_e (\phi_E - \phi_p) - D_w (\phi_p - \phi_w). \tag{5.42}$$

The continuity equation can be stated as $F_e - F_w = 0$.

To solve this discrete equation, we need to evaluate the values at the interfaces (East, West) of the control volume. Several approximation schemes have been suggested to do this. The most classical schemes in the context of meshing are presented below.

5.3.3. *Central differencing scheme*

Centered differences can be used to represent the diffusive term in the discrete equation. One seemingly reasonable approach is to attempt to approximate ϕ_e and ϕ_w by linear integration (this amounts to approximating $\frac{D\phi}{Dx}$ by centered differences):

$$\phi_e = \frac{\phi_P + \phi_E}{2}; \quad \phi_w = \frac{\phi_W + \phi_P}{2}. \qquad [5.43]$$

After substitution, we find:

$$\left[\left(D_w - \frac{F_w}{2}\right) + \left(D_e + \frac{F_e}{2}\right)\right]\phi_P = \left(D_w + \frac{F_w}{2}\right)\phi_W + \left(D_e - \frac{F_e}{2}\right)\phi_E \qquad [5.44]$$

$$\left[\left(D_w - \frac{F_w}{2}\right) + \left(D_e + \frac{F_e}{2}\right) + (F_e - F_w)\right]\phi_P = \left(D_w + \frac{F_w}{2}\right)\phi_W + \left(D_e - \frac{F_e}{2}\right)\phi_E \qquad [5.45]$$

This equation holds for any node P in the interior of the computational mesh. At the boundaries, we need a separate approach, similar to the one we used earlier for pure diffusion, to obtain a closed system. As before, this system can be written in the form:

$$a_P\phi_P = a_W\phi_W + a_E\phi_E$$

$$a_W = D_w + \frac{F_w}{2}; \quad a_E = D_e - \frac{F_e}{2}; \quad a_P = a_W + a_E + (F_e - F_w).$$

To demonstrate more clearly how the FVM works when applied to CD-type transport equations and to illustrate some of the difficulties associated with the central differencing scheme, we will consider the following CDTE:

$$\begin{cases} \frac{d}{dx}(\rho u\phi) = \frac{d}{dx}\left(\Gamma\frac{d\phi}{dx}\right) & x \in]0, L[\\ \phi(0) = \phi_0 = 1 \\ \phi(L) = \phi_L = 0 \end{cases} \qquad [5.46]$$

We will use the values $\rho = 1$ kg/m^3; $\Gamma = 0.1$ kg/m/s and $L = 10$ m, and we will consider the following cases:

1) $u = 1.0$ m/s, with a mesh of five equally-spaced points;

2) $u = 2.5$ m/s, with a mesh of five equally-spaced points;

3) $u = 2.5$ m/s, with a mesh of two equally-spaced points.

The analytic solution is:

$$\frac{\phi(x) - \phi_0}{\phi_L - \phi_0} = \frac{\exp\left(\frac{\rho u x}{\pi}\right) - 1}{\exp\left(\frac{\rho u L}{\pi}\right) - 1}. \qquad [5.47]$$

In case (2), the central differencing scheme (CDS) results in a numerical solution that oscillates around the analytic solution. This oscillation arises due to a combination of the following reasons:

Consider first of all the dimensionless quantity $P_e = \frac{uL}{\Gamma}$, known as the Péclet number (quantity proportional to the ratio of the weights of the convection and diffusion terms). If $\rho = 1$, the CDTE can be written in dimensionless form as:

$$\begin{cases} \frac{\partial \tilde{\phi}}{\partial \tilde{x}} - \frac{1}{P_e}\frac{\partial^2 \tilde{\phi}}{\partial^2 \tilde{x}} \quad x \in]0,1[\\ \tilde{\phi}(0) = 1; \quad \tilde{\phi}(1) = 0. \end{cases} \qquad [5.48]$$

This has the analytic solution:

$$\tilde{\phi}(\tilde{x}) = \frac{\exp(P_e) - \exp(P_e \tilde{x})}{\exp(P_e) - 1} \qquad [5.49]$$

– If $P_e = 0$ (pure diffusion), then:

$$\tilde{\phi}(\tilde{x}) = 1 - \tilde{x}.$$

– If $P_e \to +\infty$ (pure convection), then:

In this case, there appear to be two solutions associated with two possible choices of boundary condition. To choose the relevant solution, we need to know the direction of the current or time evolution that led to the steady state. Let h be the step size of a mesh of $N+1$ points. Discretizing the dimensionless equation by the CDS then gives:

$$\begin{cases} \frac{1}{h^2 P_e}\left(\tilde{\phi}_{i+1} - 2\tilde{\phi}_i + \tilde{\phi}_{i-1}\right) + \frac{\tilde{\phi}_{i+1} - \tilde{\phi}_{i-1}}{2h} \\ \tilde{\phi}_1 = 1; \quad \tilde{\phi}_{N+1} \end{cases} \qquad [5.50]$$

Without going into excessive detail, the general solution is:

$$\tilde{\phi}_i = a + br^i, \qquad [5.51]$$

where:

$$a = \frac{-r^N}{1 - r^N}; \quad b = \frac{1}{r - r^{N+1}}; \quad r = \frac{2 + hP_e}{2 - hP_e}.$$

Therefore, the behavior of the solution fundamentally depends on the sign of r, via the Péclet term with mesh P_{en} ($P_{en} = hP_e$), which reflects the fineness of the discretization.

In summary:

– If $P_{en} > 2$, then the mesh is too coarse for the given values of u and Γ. The numerical solution will exhibit sawtooth oscillations (every other node on the mesh).

– If $P_{en} < 2$, then the mesh is appropriately scaled relative to the phenomena being modeled. The numerical solution is monotone and converges to the exact solution at a rate of h^2. This extremely simple example illustrates the limitations of the second-order approach (CDS). If we can satisfy the constraint on the Péclet number (which may be possible in some cases, for example, at very low speeds), then this approach can be viable, because it achieves a high order of convergence (second order in this case) with a simple discretization. However, if the mesh is too coarse in regions of the domain with high gradients, then an oscillatory zone can develop and spread to the rest of the domain, rendering the numerical solution physically useless. Instead, less "heavy" approximations are often preferable, motivated by the observation that a quantity defined to be positive can otherwise take negative values.

This discussion suggests that discretization schemes should satisfy certain fundamental properties in order to be exploitable in practice for scientific computations, especially in the context of finite difference methods. Among the most important of these desirable properties, we can quote the following: a good numerical scheme should be conservative, bounded and transportive.

5.3.4. *Upwind (decentered) scheme*

We saw earlier that the formulation of the CDS does not incorporate the direction of the flow, i.e. the value ϕ_w is always influenced by both ϕ_p and ϕ_W.

If there is strong convection from West to East, the CDS ceases to be viable, because ϕ_w needs to be more strongly influenced by ϕ_W than by ϕ_p. The upwind scheme was introduced to model the direction of flow when approximating ϕ at the interfaces (w, e) (see Figure 5.7).

– If the flow is in the positive direction:

$$u_w > 0, \quad u_e > 0 \quad (F_e, F_w > 0),$$

then the upwind scheme approximates ϕ_p and ϕ_w by:

$$\phi_e = \phi_p \quad \phi_w = \phi_W.$$

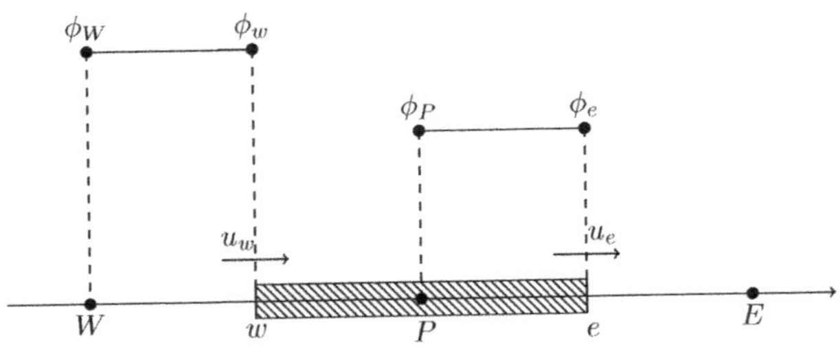

Figure 5.7. *Upwind scheme (positive direction)*

The discrete CDTE then becomes:

$$[(D_w + F_w) + D_e + (F_e - F_w)]\,\phi_P = (D_w + F_w)\,\phi_W + D_e\phi_E. \qquad [5.52]$$

– If the flow is in the negative direction:

$$u_w < 0, \quad u_e < 0 \quad (F_e, F_w < 0),$$

then the upwind scheme approximates ϕ_p and ϕ_w by

$$\phi_e = \phi_E \quad \phi_w = \phi_P.$$

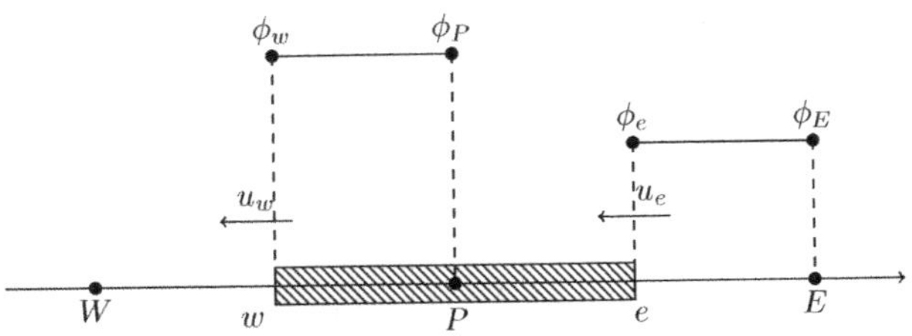

Figure 5.8. *Upwind scheme (negative direction)*

The corresponding discrete CDTE is:

$$[D_w + (D_e - F_e) + (F_e - F_w)]\,\phi_p = D_w\phi_W + (D_e F_e)\,\phi_E. \qquad [5.53]$$

In summary, if we write the general form of the discrete CDTE as:

$$a_p\phi_p = a_E\phi_E + a_W\phi_W, \qquad [5.54]$$

then the coefficients a_p, a_E and a_W are defined as:

$$a_w = D_w + max(F_w, 0) \qquad [5.55]$$

$$a_E = D_e + max(F_w, 0) \qquad [5.56]$$

$$a_p = a_E + a_W + (F_e - F_w). \qquad [5.57]$$

EXAMPLE.– Consider again cases (1) and (2) from the previous example. The upwind scheme (US) is conservative, because it uses consistent expressions to compute the flux terms through each face of the control volume.

It is also bounded, because the coefficients a_E, a_p and a_W are always positive. Furthermore, when the continuity equation is satisfied $(F_e - F_w = 0)$, $a_p = a_E + a_W$; in other words, the system matrix is diagonally dominant, which prevents oscillations from arising. Finally, the upwind scheme is transportive, by construction. It is worth noting that the upwind scheme does have one disadvantage: its precision (rate of convergence). This scheme uses first-order formulas to approximate the gradients of ϕ, which introduces a numerical diffusion term (unrelated to any physical phenomenon) that degrades the numerical solution. In other words, the decentering model comes hand-in-hand with an artificial diffusion term that depends on the speed and the discretization step. In general, the decentering can be expressed as:

$$\frac{\partial \phi}{\partial x} = \alpha\frac{\phi_W - \phi_P}{h} + (1-\alpha)\frac{\phi_E - \phi_P}{h} + (\alpha - 1/2)\,o(h) \qquad [5.58]$$

– upwind decentering: $a > 1/2$

– downwind decentering: $a < 1/2$

$$u\frac{\partial \phi}{\partial x} = \alpha u\frac{\phi_W - \phi_P}{h} + (1-\alpha)u\frac{\phi_E - \phi_P}{h} + o(h). \qquad [5.59]$$

This relation can be approximated to the second order as follows (for comparison with the centered approximation of the diffusion term):

$$u\frac{\phi_E - \phi_P}{h} + (1/2-\alpha)uh\frac{\phi_E - 2\phi_p + \phi_W}{h^2} = u\frac{\partial \phi}{\partial x} - (\alpha - 1/2)uh\frac{\partial \phi^2}{\partial x^2} + o(h^2). [5.60]$$

Therefore, by decentering in the direction of the current ($Q > 2$), we use the second-order expression, but the problem is stabilized by formally adding an artificial diffusion of the order of $D_{num} = (\alpha - 1/2)|u|h$.

5.3.5. *Hybrid scheme*

When the speed and the diffusion coefficients of the flow vary over time and space (which is often the case in practice), the Péclet number also varies; there may be some zones with low Péclet numbers and others with high Péclet numbers. One approach to designing a numerical scheme for this type of situation is to define a "hybrid" scheme (HS), for example, by approximating q_w with the following definition:

$$\begin{cases} q_w = F_w\left[\frac{1}{2}\left(1+\frac{2}{P_{ew}}\right)\phi_W + \frac{1}{2}\left(1-\frac{2}{P_{ew}}\right)\phi_P\right] & \text{if } -2 < P_{ew} < 2 \\ q_w = F_w A_1 \phi_W & \text{if } P_{ew} \geq 2 \\ q_w = F_w A_2 \phi_P & \text{if } P_{ew} \leq 2 \end{cases} \quad [5.61]$$

where:

$$P_{ew} = \frac{F_w}{D_w} = \frac{(\rho u)_w}{\Gamma_w/\delta_{WP}}.$$

In summary, the coefficients of this discretization scheme, $a_P\phi_P = a_E\phi_E = a_W\phi_W$, are as follows:

$$\begin{cases} a_W = \max\left[F_w; \left(D_w + \frac{F_w}{2}\right); 0\right] \\ a_E = \max\left[-F_e; \left(D_e + \frac{F_e}{2}\right); 0\right] \\ a_p = a_E + a_W \end{cases} \quad [5.62]$$

5.3.6. *Power-law scheme*

Approximations using the power-law scheme are more precise in one dimension and produce better results than the hybrid scheme. The power-law scheme neglects the effects of diffusion when the Péclet number is greater than 10. If P_e is between 0 and 10, the flux is evaluated with a polynomial expression. For example, the flux per unit surface area at w is:

$$\begin{cases} q_w = F_w\left[\phi_W - \beta_w\left(\phi_P - \phi_W\right)\right] & \text{if } 0 < P_e < 10 \\ q_w = F_w\phi_W & \text{if } P_e \geq 10 \end{cases} \quad [5.63]$$

where $\beta_w = \dfrac{\left(1 - \dfrac{P_{ew}}{10}\right)^5}{P_{ew}}$.

The coefficients a_p, a_E and a_w of this method are defined as follows:

$$\begin{cases} a_E = D_e \max\left[0; \left(1 - \left(\dfrac{|P_{ew}|}{10}\right)^5\right)\right] + \max[F_e; 0] \\ a_W = D_w \max\left[0; \left(1 - \left(\dfrac{|P_{ew}|}{10}\right)^5\right)\right] + \max[F_w; 0] \\ a_p = a_E + a_W \end{cases} \qquad [5.64]$$

This scheme is very widely used in practice as an alternative to the hybrid scheme. The commercial software package FLUENT v4.22 uses it as the default scheme for computing the flow.

5.3.7. *QUICK scheme*

QUICK (Quadratic Upstream Interpolation for Convection Kinetics) uses quadratic interpolation on three nodes to compute the value at the interface. In the direction of flow, we consider two upstream nodes and one downstream node. For example:

– For $u_w > 0$; $u_e > 0$, we use quadratic smoothing between WW, W and P to evaluate ϕ_w, and a separate instance of smoothing between W, P and E to evaluate ϕ_e (see Figure 5.9).

– For $u_w < 0$; $u_e < 0$, the values of b at W, P and E are used to evaluate ϕ_w, and the values of ϕ at EE, P and E are used to evaluate ϕ_e:

$$\begin{cases} u_w > 0, \phi_w = \dfrac{6}{8}\phi_W + \dfrac{3}{8}\phi_P - \dfrac{1}{8}\phi_W W \\ u_e > 0, \phi_e = \dfrac{6}{8}\phi_P + \dfrac{3}{8}\phi_E - \dfrac{1}{8}\phi_W \\ u_w < 0, \phi_w = \dfrac{6}{8}\phi_P + \dfrac{3}{8}\phi_W - \dfrac{1}{8}\phi_E \\ u_e < 0, \phi_e = \dfrac{6}{8}\phi_E + \dfrac{3}{8}\phi_P - \dfrac{1}{8}\phi_E E \end{cases} \qquad [5.65]$$

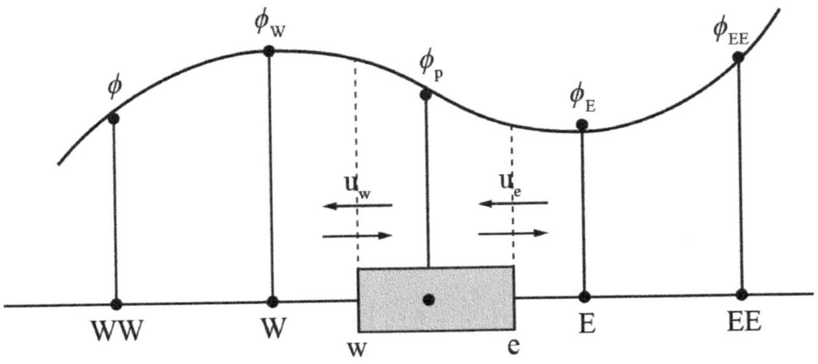

Figure 5.9. *Quadratic smoothing*

Note that the discrete CDTE includes two additional coefficients a_{WW} and a_{EE}. This is because we used ϕ_{EE} to approximate the flux at the interfaces. The expressions of a_p, a_E, a_W, a_{WW} and a_{EE} can be stated more concisely as follows:

$$\begin{cases} a_W = D_w + \frac{6}{8}\alpha_w F_w + \frac{1}{8}\alpha_e F_e - \frac{3}{8}(1-\alpha_w)F_w \\ a_E = D_e - \frac{3}{8}\alpha_e F_e - \frac{6}{8}(1-\alpha_e)F_e - \frac{1}{8}(1-\alpha_w)F_w \\ a_{EE} = \frac{1}{8}(1-\alpha_e)F_e \\ a_{WW} = -\frac{1}{8}\alpha_w F_w \\ a_p = e_E + a_W + a_{EE} + a_{WW} + (F_e - F_w) \\ a_p\phi_p = e_E\phi_E + a_W\phi_W + a_{EE}\phi_{EE} + a_{WW}\phi_{EE} \end{cases} \quad [5.66]$$

where:

$$\alpha_{w,e} = \begin{cases} 1 & \text{if } u_{w,e} > 0 \\ 0 & \text{if } u_{w,e} < 0 \end{cases} \quad [5.67]$$

The flux terms at the interfaces are calculated by quadratic interpolation on two upstream nodes and one downstream node. This procedure guarantees that the QUICK scheme is conservative.

QUICK is accurate up to the third order on uniform grids. The "transportativity" criterion is satisfied by construction as a quadratic function that uses the values of ϕ at two upstream nodes and one downstream node. If the flow also satisfies the continuity equation, then the coefficient a_p is the sum of the neighboring coefficients. This is a desirable property when trying to show that the scheme is bounded. However, it can easily be seen that a_E and a_W are not always positive for the QUICK scheme, and a_{WW}, a_{EE} are negative. Therefore, QUICK is not always stable; we say that it is conditionally stable. Another disadvantage of this scheme is that the structure of the linear system matrix that we must solve is pentadiagonal, unlike the schemes considered above, which produce tridiagonal matrices that are easier and quicker to solve.

Finally, we note that special care must be taken with QUICK when discretizing the boundaries of the domain, because fictitious points (mirror points) are required to close the system. To resolve this problem, we typically use extrapolation of the form:

$$\phi_A = \frac{\phi_0 + \phi_1}{2} \quad \phi_0 = 2\phi_A - \phi_1. \qquad [5.68]$$

5.3.8. *Higher-order schemes*

Even though QUICK is accurate to the third order, it can still develop undesirable oscillations. This motivated the development of second-order schemes without parasitic oscillatory behavior. Two examples of such schemes are "slope limiters" and total variation diminishing (TVD) schemes. For example, slope-limiter schemes model the flux at the interface "e" by:

$$\phi_e = \begin{cases} \phi_p + \frac{1}{2}(\phi_E - \phi_p)\,\psi(\theta_e^+)\ if\ u_e \geq 0 \\ \\ \phi_E - \frac{1}{2}(\phi_E - \phi_p)\,\psi(\theta_e^-)\ if\ u_e < 0 \end{cases} \qquad [5.69]$$

The function ψ is called the limiter function, and is designed to stabilize the scheme and eliminate parasitic oscillations.

The best-known examples of limiters are:

– Minmod limiter: $\phi(\theta) = \max(0, \min(1, \theta))$;

– Superbee limiter: $\phi(\theta) = \max(0, \min(1, 2\theta), \min(2, \theta))$;

– van Leer limiter: $\phi(\theta) = (\theta + |\theta|)/(1 + |\theta|)$;

– MC limiter: $\phi(\theta) = \max[0, \min((1+\theta)/2, 2\theta)]$.

Each of these limiters satisfies the stability condition:

$$0 \leqslant \frac{\phi(\theta)}{\theta} \leqslant 2 \quad \text{and} \quad 0 \leqslant \phi(\theta) \leqslant 2. \qquad [5.70]$$

These schemes limit the numerical diffusion and eliminate oscillations while remaining second-order overall.

5.3.9. *Unsteady one-dimensional convection-diffusion equation*

In this section, we present an example of a time-dependent flow problem by considering the following transport equation:

$$\frac{\partial \phi}{\partial t} + \text{div}\,(\rho \vec{u} \phi) = \text{div}\left(\Gamma \vec{\text{grad}} \phi\right) + S_\phi. \qquad [5.71]$$

When applying the FVM to transient problems, in addition to integrating the equation over control volumes as usual, we need to integrate the equation over a time interval of size Δt :

$$\int_{\Delta v}\left(\int_t^{t+\Delta t} \frac{\partial}{\partial t}(\rho\phi)dt\right)dv + \int_t^{t+\Delta t}\left(\int_A \vec{n}.(\rho\phi\vec{u})dA\right)dt = \int_t^{t+\Delta t}\left(\int_A \vec{n}.(\Gamma \vec{\text{grad}}\phi)dA\right)dt + \int_t^{t+\Delta t}\left(\int_{\Delta v} S_\phi dv\right)dt. \qquad [5.72]$$

The convection-diffusion terms and the time-based terms need to be approximated simultaneously.

The methods for approximating the convection-diffusion terms presented above (in the steady-state case) can be reused for the transient case (under certain conditions). To demonstrate the method used for integration with respect to time and without loss of generality, we will consider the unsteady pure diffusion equation (heat conduction equation):

$$\rho c \frac{\partial T}{\partial t} = \frac{\partial}{\partial x}\left(k \frac{\partial T}{\partial x}\right) + S. \qquad [5.73]$$

Integrating this equation over a control volume and $[t, t+\Delta t]$ gives:

$$\int_w^e\left(\int_t^{t+\Delta t} \rho c \frac{\partial T}{\partial t} dt\right)dv = \int_t^{t+\Delta t}\left(\int_A \vec{n}.(\rho\phi\vec{u})dA\right)dt = \int_t^{t+\Delta t}\left[\left(kA\frac{\partial T}{\partial x}\right)_e - \left(kA\frac{\partial T}{\partial x}\right)_w\right]dt + \int_t^{t+\Delta t} \bar{S}\Delta v dt. \qquad [5.74]$$

The left-hand term can be approximated by:

$$\int_{\Delta v}\left(\int_{t}^{t+\Delta t}\rho c\frac{\partial T}{\partial t}dt\right)dv=\rho c(T_p-T_p^{\circ})\Delta v, \qquad [5.75]$$

where T_p° denotes the temperature at the point P at time t and T_p denotes the temperature at the point P at time $t+\Delta t$.

The same result can be obtained by replacing $\frac{\partial T}{\partial t}$ by $\frac{T_p-T_p^{\circ}}{\Delta t}$, i.e. by approximating $\frac{\partial T}{\partial t}$ using a first-order decentering scheme. The discrete equation can be stated as:

$$\rho c(T_p-T_p^{\circ})\Delta v=\int_{t}^{t+\Delta t}\left[\left(k_eA_e\frac{T_E-T_p}{\delta_{EP}}\right)\right.$$
$$\left.-\left(k_wA_w\frac{T_p-T_W}{\delta_{pW}}\right)\right]+\int_{t}^{t+\Delta t}\bar{S}\Delta v dt. \qquad [5.76]$$

We must now decide which values of T_w, T_P and T_E we should use (at t or $t+\Delta t$) to approximate the first integral above. Depending on our choice, we obtain two types of time discretization: explicit (using the values at t) and implicit (using the values at $t+\Delta t$).

Both choices can be generalized into a θ-scheme with the following general expression:

$$I_p=\int_{t}^{t+\Delta t}T_pdt=\left[\theta T_p+(1-\theta)\,T_p^{o}\right]\Delta t. \qquad [5.77]$$

Thus:

– for $\theta=0$, we recover the explicit scheme;

– for $\theta=1$, we recover the implicit scheme;

– for $\theta=\frac{1}{2}$, we recover the Crank–Nicolson scheme.

With the θ-scheme, the discrete equation is:

$$a_pT_p=a_W\left[\theta T_W+(1-\theta)\,T_W^{o}\right]+a_E\left[\theta T_E+(1-\theta)\,T_E^{o}\right]$$
$$+\left[a_p^{o}-(1-\theta)\,a_W-(1-\theta)\,a_E\right]T_p^{o}+b, \qquad [5.78]$$

where:

$$\begin{cases} a_p = \theta\left(a_W + a_E\right) + a_p^o \\ a_p^o = \rho c \dfrac{\Delta x}{\Delta t} \\ b = \bar{S}\Delta x \end{cases} \qquad [5.79]$$

5.3.10. *Explicit scheme*

If we linearize the source term $b = S_u + S_p T_p^{\circ}$ and substitute $\theta = 0$ into the above equation, we obtain the "explicit" discretization of the unsteady heat conduction equation:

$$a_p T_p = a_W T_W^o + a_E T_E^o + \left[a_p^o - (a_W + a_E - S_p)\right] T_p^o + S_u \qquad [5.80]$$

$$a_p = a_p^o = \rho c \frac{\Delta x}{\Delta t}; \quad a_W = \frac{k_w}{\delta x_{Wp}}; \quad a_E = \frac{k_e}{\delta x_{Ep}}.$$

We note that T_p can be computed explicitly from the value of T at time t, without needing to solve a linear system.

For reasons of stability, the coefficients of the equation must be positive. This is in particular true for the coefficient of T_p^o; in other words, the relation $a_p - a_W - a_E > 0$ must hold.

In the case of a regular mesh $\Delta x = \delta x_{Wp} = \delta x_{pE}$ and constant k, the above condition can be rewritten as:

$$\rho c \frac{\Delta x}{\Delta t} > \frac{2k}{\Delta x} \Longrightarrow \Delta t < \rho c \frac{\Delta x^2}{2k}. \qquad [5.81]$$

This condition specifies the maximum time step before the explicit scheme fails. It represents a limitation of the discretization approach, in the sense that improving the spatial precision (Δx small) penalizes the computation time (the limiting Δt becomes small). Nonetheless, with a sensible choice of Δt, explicit discretization is viable for heat condition problems.

5.3.11. *Crank–Nicolson scheme*

This scheme chooses $\theta = \frac{1}{2}$ in the discrete equation:

$$a_p T_p = a_E \left(\frac{T_E + T_E^o}{2}\right) + a_W \left(\frac{T_W + T_W^o}{2}\right) + \left[a_p^o - \frac{a_E}{2} - \frac{a_W}{2}\right] T_p^o + b, \quad [5.82]$$

where:

$$\begin{cases} a_p = \dfrac{1}{2}(a_W + a_E) + a_p^o - \dfrac{1}{2}S_p \\ a_p^o = \rho c\dfrac{\Delta x}{\Delta t} \\ b \;= S_u + \dfrac{1}{2}S_pT_p^o \end{cases} \qquad [5.83]$$

We note that this scheme requires a linear system to be solved at every time step. It is an implicit scheme that is unconditionally stable. To guarantee the positivity of the coefficients, the relation $a_p^o > \frac{a_E+a_W}{2}$ must hold, leading to the condition:

$$\Delta t < \rho c\frac{\Delta x^2}{k}. \qquad [5.84]$$

This condition is less restrictive than the condition imposed by the explicit scheme. The Crank–Nicolson scheme uses a centered difference for $\frac{\partial T}{\partial t}$, and is therefore second-order in time. With a sufficiently small step size Δt, it can produce solutions with higher overall accuracy than the explicit scheme. In fact, the overall accuracy of the computations depends on the spatial discretization. Therefore, the Crank–Nicolson scheme is typically used with spatially centered differences.

5.3.12. *Implicit scheme*

The implicit scheme is obtained by choosing $\theta = 1$:

$$a_pT_p = a_WT_W + a_ET_E + a_p^oT_p + S_u, \qquad [5.85]$$

where:

$$a_p = a_W + a_E + a_p^o - S_p; a_p = a_p^o = \rho c\frac{\Delta x}{\Delta t};$$

$$a_W = \frac{k_w}{\delta x_{Wp}}; a_E = \frac{k_e}{\delta x_{Ep}}.$$

The discrete equation [5.85] incorporates the temperature values at the neighbors of the node P, which are unknown, and are therefore determined implicitly by solving a linear system (hence the name "implicit scheme"). Every coefficient of the resulting linear system is positive. The implicit scheme is therefore unconditionally stable. It is first order in time, because it uses a decentered time discretization. To guarantee a certain level of precision in time, a suitable value of Δt must be chosen, while bearing in mind the associated computational costs. The implicit scheme is the most commonly used approach for unsteady meshing problems because it is straightforward to implement and is unconditionally stable.

5.4. Using *Matlab*

We will solve the conservative form of the shallow water equation in two dimensions using the finite volume method in *Matlab*. The corresponding script is provided below:

```
clear all; clc;
% Construct the grid
m = 60;
dx = 2/m; dy = 2/m;
x = -1-dx:dx:1+dx; y = -1-dy:dy:1+dy;
[xx,yy] = meshgrid(x,y);
g = 1; c = 0.5;

h = ones(size(xx));
h(xx>=-0.5 & xx<=0.5 & yy>=-0.5 & yy<=0.5) = 2;
% U is an unknown matrix.
 U(:,:,1)=h,U(:,:2)=hu,U(:,:,3)=hv
U = zeros([size(h) 3]);
U(:,:,1) = h;
u = zeros(size(xx));
v = u;

% circular shift vectors
shiftp1 = circshift((1:length(x))',1);

shiftm1 = circshift((1:length(x))',-1);

mesh(x,y,U(:,:,1)), colormap jet, axis([-1 1 -1 1 0.5 2.5])
title('hit enter to continue')
xlabel x, ylabel y; zlabel h;
pause;

t = 0; dt = 0; tstop = 3.0;
ii = 1;
numplots = 3;
```

```
tplot = [1.35;3.0]; Uplot = zeros([size(U) length(tplot)+1]);
Uplot(:,:,:,1) = U; styles = {'k:','k--','k-'};

while t < tstop
    Uold = U; uold = u; vold = v; told = t; t = t + dt;

    % calculate lambda = |u| + sqrt(gh) using the computed value
       of the flux
    lambdau = 0.5*abs(uold+uold(:,shiftm1)) +...
        sqrt(g*0.5*(Uold(:,:,1)+Uold(:,shiftm1,1)));
    lambdav = 0.5*abs(vold+vold(shiftm1,:)) +...
        sqrt(g*0.5*(Uold(:,:,1)+Uold(shiftm1,:,1)));
    lambdamax = norm([lambdau(:); lambdav(:)],Inf);

    dt = c*(dx/lambdamax);
    % adjust dt to regenerate the graph
    if (ii<=length(tplot) && tplot(ii)>=told && tplot(ii)<=t+dt)
        dt = tplot(ii)-t;
        ii = ii + 1;
    end

    huv = Uold(:,:,2).*Uold(:,:,3)./Uold(:,:,1);
    ghh = 0.5*g*Uold(:,:,1).^2;
    % compute (hu,hu^2+gh^2/2,huv)
    lffu = cat(3,Uold(:,:,2),Uold(:,:,2).^2./Uold(:,:,1)+ghh,huv);
    % compute (hv,huv,hv^2+gh^2/2)
    lffv = cat(3,Uold(:,:,3),huv,Uold(:,:,3).^2./Uold(:,:,1)+ghh);
    % compute the flux terms
    fluxx =  0.5*(lffu+lffu(:,shiftm1,:)) - ...
       0.5*bsxfun(@times,Uold(:,shiftm1,:)-Uold,lambdau);
    fluxy =  0.5*(lffv+lffv(shiftm1,:,:)) - ...
       0.5*bsxfun(@times,Uold(shiftm1,:,:)-Uold,lambdav);
    % time step
    U = Uold - (dt/dx)*(fluxx - fluxx(:,shiftp1,:)) ...
       - (dt/dy)*(fluxy - fluxy(shiftp1,:,:));

    % impose the boundary conditions on h
    U(1:end,end,1)= U(1:end,end-1,1); U(1:end,1,1)= U(1:end,2,1);
    U(end,1:end,1)= U(end-1,1:end,1); U(1,1:end,1)= U(2,1:end,1);
    % on hu
    U(1:end,end,2)=-U(1:end,end-1,2); U(1:end,1,2)=-U(1:end,2,2);
    U(end,1:end,2)= U(end-1,1:end,2); U(1,1:end,2)= U(2,1:end,2);
    % on hv
```

```
    U(1:end,end,3)= U(1:end,end-1,3); U(1:end,1,3)=U(1:end,2,3);
    U(end,1:end,3)=-U(end-1,1:end,3); U(1,1:end,3)=-U(2,1:end,3);

    % u = hu./h;                  % v = hv./h;
    u = U(:,:,2)./U(:,:,1);       v = U(:,:,3)./U(:,:,1);

    % display the animation
    mesh(x,y,U(:,:,1)), colormap jet, axis([-1 1 -1 1 0 2.5])
    title(['t = ' num2str(t+dt)])
    xlabel x, ylabel y, zlabel y, pause(0.001)
    %if (ismember(t+dt,tplot))
    if (any(tplot-t-dt==0))
        Uplot(:,:,:,ii) = U;        % store U for plotting
    end
end
```

The result returned by this script is shown in Figure 5.10.

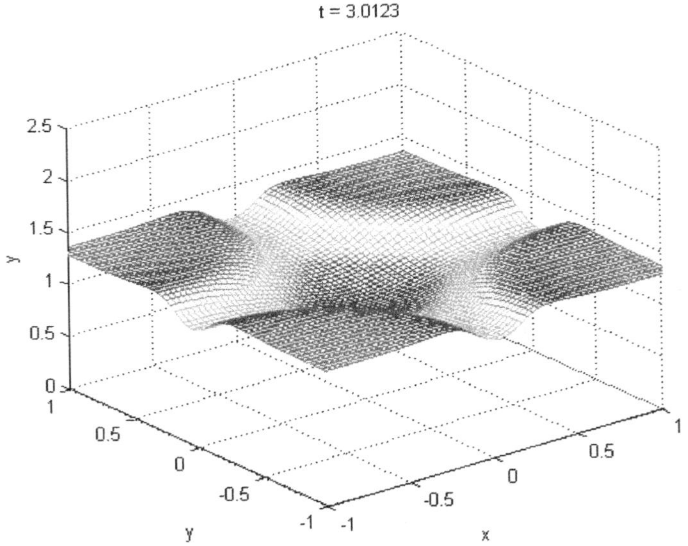

Figure 5.10. *Solution of the shallow water equation computed by finite volumes. For a color version of this figure, see www.iste.co.uk/radi/advanced2.zip*

6

Meshless Methods

6.1. Introduction

Meshless methods offer significant advantages in nonlinear structure analysis because they can handle discontinuities and large deformations flexibly, whereas FEM approaches typically suffer from severe grid distortion under these conditions. Research into structure elastoplasticity using meshless methods has remained a relatively niche field in the past, and has only begun to attract attention more recently.

Chen *et al.* [CHE 96] presented a formulation for nonlinear structures undergoing large deformations based on the reproducing kernel particle method (RKPM) and used it to study elastoplasticity and hyperelasticity problems. Rao and Rahman [RAO 04] proposed an augmented meshless method for analyzing nonlinear ruptures to study the fissuring of solids. Kargarnovin *et al.* [KAR 04] extended the element-free Galerkin (EFG) method to elastoplasticity stress analysis using an incremental formulation of plastic deformations. Xu and Saigal [XU 98, XU 99] suggested an EFG-based approach to describe the propagation of quasi-static and dynamically stable fissures in elastoplastic materials.

Liu *et al.* [LIU 05] used the EFG method coupled with FEM to solve elastoplastic contact problems. Belinha and Dinis [BEL 06, BEL 07] conducted elastoplastic analysis of plates using the EFG method. This method was also used by other researchers to perform nonlinear analysis of folded plates [LIE 07], analyze the elastoplastic adaptation of structures made from perfectly plastic materials [CHE 08] and simulate nonlinear dynamic ruptures [RAB 07]. Other fields of application of meshless methods include the analysis of unsteady systems [CHE 09].

6.2. Limitations of the FEM and motivation of meshless methods

Numerical simulations have become truly irreplaceable for certain types of process, while other processes have remained unexplored. In particular, anything involving very large deformations, high deformation speeds, separation of matter or extremely localized deformations often requires the complex differential equations governing these phenomena to be solved. In general, differential equations are usually solved by numerical methods such as the finite element method (FEM), the finite difference method (FDM) or the finite volume method (FVM). By using a suitably predefined mesh and applying the principles of these methods to the problem, even complex differential equations can be approximated by systems of equations that are easier to solve.

The finite element method (FEM) [ZIE 00] is now the most widely used approach for solving the partial differential equations of physical and mechanical systems. However, it suffers from some limitations during simulations. The necessity of reconstructing the mesh, for example, when large transformations are encountered – to either avoid deformation of the elements or match the shape of the mesh to the localizations – creates significant extra computational cost, as well as introducing robustness issues, especially when working with complex three-dimensional geometries.

Over the last two decades, a range of new numerical methods have been developed as alternatives to the finite element method. Many of these methods circumvent the difficulties associated with the mesh by constructing some or all of the approximations using techniques that do not rely on spatial discretization elements. “Meshless methods” are one such group of techniques. They have proven to be effective for certain problems that are difficult to solve using the finite element method.

6.3. Examples of meshless methods

The shape functions of meshless methods are not constructed by partitioning the domain into elements, but are instead defined only in terms of a cloud of nodes. The first such method was proposed in the 1970s. A few examples are listed below:

– the smooth particle hydrodynamics (SPH) method [LUC 77], which simulates astrophysical phenomena such as star explosions using a set of particles;

– the diffuse element method (DEM) [NAY 92], which uses a basis function and a set of weighted nodes to construct an approximation of the displacement field;

– the element-free Galerkin (EFG) method [BEL 94], which computes the moving least-squares approximation (MLSA) from the previous approximation [LAN 81];

– the reproducing kernel particle method [LIU 05];

– the hp-clouds method [DUA 96];

– the partition of unity finite element method [MEL 96], which uses FE shape functions and polynomial basis functions;

– the boundary node method [MUK 97];

– the meshless local Petrov–Galerkin (MLPG) method [ATL 98].

In this chapter, we will study and apply the element-free Galerkin method proposed by Belytschko, Lu and Gu.

6.3.1. *Advantages of meshless methods*

The main advantage of using meshless methods in simulations is that they allow problems with large transformations to be handled more easily than the finite element method. Their excellent performance under these conditions can be traced back to the following factors:

1) In Lagrangian formulations, which are total, the gradient operator of the transformation computed at a given integration point is constructed from a neighborhood of nodes that is typically larger than just the nodes of the element considered by the FEM. Therefore, the distortions in the neighborhood can be much larger before the Jacobian matrix becomes singular.

2) The quality of the solution is much less sensitive to the relative positions of the nodes. In updated Lagrangian-type formulations, this allows us to construct the solution from these relative positions, which is not viable in the context of the finite element method.

3) The fact that we do not need to construct a mesh to build the approximation allows us to handle domains with complex 2D and 3D geometries with just a cloud of nodes.

Another major advantage of meshless methods is that they make it extremely easy to insert or remove specific nodes, because the relative positions between the nodes have very little effect on the quality of the solution.

Meshless shape functions are generally highly isotropic, which significantly reduces issues associated with dependencies between the directions of the shear bands or fissures and the mesh. This property was illustrated by successfully simulating the propagation of the shear bands with the RKPM.

Finally, the relatively wide support of the moving-least-squares shape functions somewhat mitigates the blockage issues encountered in incompressibility problems.

For a while, it was believed that meshless methods were completely exempt from these blockage problems, but this has since been disproven. Whenever narrower supports are used for the shape functions (for reasons of cost and/or quality), the blockage problems resurface. Thus, although less significant, blockage remains a concern in the general case.

6.3.2. *Disadvantages of meshless methods*

The greatest disadvantage of most "classical" meshless methods (SPH, RKPM, DEM, EFG) is the difficulty in imposing boundary conditions. In order to be able to impose Dirichlet-type boundary conditions directly, like in the finite element method, the approximation must be constructed in terms of the node values (strict interpolation) and the influence of interior nodes must cancel on the boundary of the domain. However, the approximation functions constructed by the most common meshless methods do not satisfy either of these conditions. This problem has inspired a large number of publications to propose a range of techniques for imposing boundary conditions in meshless methods, such as the Lagrange multiplier method [BEL 94], the transformation method [CHE 96], an approach based on d'Alembert's principle [GUN 97], the introduction of singular weight functions [KAL 97], a penalty-based method [ZHU 98], and a mixed transformation method [CHE 08] to name a few. More recently, [CHE 09] has proposed a technique that allows the RKPM shape functions to act as interpolants. Although these techniques are effective, they introduce additional costs, and make it difficult to use updated Lagrangian-type approaches, which apply the boundary conditions to the updated configuration. Another rapidly abandoned technique was to couple a meshless method to a layer of finite elements at the boundary of the domain, discretizing the interior of the domain by the meshless approach [ATT 94]. This technique naturally caused all of the usual mesh-related problems to resurface.

The second disadvantage of meshless methods is linked to numerical integration. In most cases, meshless shape functions have rational rather than polynomial expressions, which renders Gaussian-type integration schemes non-optimal. Dolbow and Belytschko [DOL 99] proved that extremely fine integration schemes are required to minimize the error caused by the non-overlapping of the support of the shape functions and the integration cells, requiring very large numbers of integration points and hence excessive computational costs. Conversely, Chen et al. [CHE 98] showed that direct nodal integration leads to numerical instability and therefore cannot be used. Various solutions to this obstacle have since been proposed.

Another issue relates to the support of the shape functions. In most meshless approaches, the support or domain of influence of each node is defined by a sphere or a parallelepiped centered around the node. As discussed by Liu *et al.* [LIU 95, LIU 96], this support must include sufficiently many particles for the method to be stable and therefore must be sufficiently wide. However, excessively large support leads to high computational costs and strongly degrades the quality of the solution. Necessary conditions for guaranteeing the stability of these methods and ensuring that the basis functions are properly reproduced are stated in [LIU 95]. Using fixed supports in problems in which the node cloud experiences strong distortions can cause instability to the method. Continuously readjusting the size of the support during the simulation can help to avoid this problem, but creates robustness problems, because choosing the suitable support size relative to the local density is a non-trivial problem. This significantly limits the applicability of these methods in adaptive refinement problems, in which the nodal density is strongly heterogeneous across distinct regions in the domain, as well as problems involving separation of matter, such as machining simulations.

6.3.3. *Comparison of the finite element method and meshless methods*

Similar to the finite element method, the meshless approach solves the weak form of the PDE using a Galerkin method; however, the approximation of the displacement field in the weak formulation does not require a mesh. Instead, a set of nodes is distributed throughout the domain, and the approximation of the displacement field at any given point only depends on the distance of this point from its neighbors, and not on whether it belongs to any specific finite element. Thus, meshless methods only differ from the finite element method in some regards; both approaches adopt a similar solving structure or procedure. The most significant difference is how the interpolation functions are computed by the FEM; the meshless approach uses shape functions because it does not have a notion of element.

6.4. Basis of meshless methods

6.4.1. *Approximations*

Meshless methods use the following approximation to describe the scalar function u in terms of the (Lagrangian) coordinates of the material:

$$u_h(x,t) = \sum_{i \in S} \phi_i(x) u_i(t), \qquad [6.1]$$

where $\phi_i\colon \Omega \to \mathbb{R}$ are the shape functions, u_i is the value at the i-th node, which is located at position x_i, and S is the set of nodes i such that $\phi_i(x) \neq 0$. We note that this formula is identical to the approximation used by the FEM. However, unlike the FEM, the shape functions in equation [6.1] are simply approximations and not interpolations because $u_i \neq u(x_i)$. Consequently, special techniques are required to implement boundary conditions that are phrased in terms of the displacement. These techniques are discussed further below.

6.4.2. *Kernel (weight) functions*

The shape functions ϕ_i are derived from a set of kernel functions, also known as weight functions, denoted $\omega_i : \Omega \to \mathbb{R}$. These kernel functions have compact support, whose size is determined by the choice of dilatation parameter. This parameter plays a crucial role in both the exactness of the solution and its stability, and is roughly analogous to the size of the elements in the finite element method.

Finally, each weight function has a certain shape, required to be continuous and positive on its support. In every meshless method discussed below, the continuity of the shape functions only depends on the continuity of the kernel functions; for more details, see [HUE 04]. For example, if the kernel function is C^2, then the corresponding shape function will also be C^2.

6.4.3. *Completeness*

Completeness, often linked to reproducibility, plays a role in Galerkin methods analogous to that of consistency in the finite difference method. Completeness refers to the capacity of an approximation to reproduce polynomials of a certain order. An approximation is said to be complete to order zero if it reproduces constant functions exactly. It is said to be linear (complete to first order) if it reproduces linear functions exactly and so on for higher orders.

6.4.4. *Partition of unity*

Partition of unity (PU) refers to the division of the domain into overlapping subdomains Ω_i, each associated with a function $\phi_i(x)$ that is zero everywhere except on Ω_i, and which furthermore satisfies the following property:

$$\sum_{i=1}^{N} \phi_i(x) = 1 \quad \text{on} \quad \Omega. \qquad [6.2]$$

There are two ways of increasing the completeness order of the approximation. The first is to intrinsically increase the completeness order of the shape functions by directly increasing the completeness order of the kernel functions. Alternatively, the completeness order can be increased by modifying equation [6.1] according to the principle of partition of unity (PU).

6.5. Meshless method (EFG)

6.5.1. *Theory*

The EFG method is classified as meshless because it only requires a set of nodes and a description of the boundary to construct an approximate solution. The contrast between the meshless approach and the method of finite elements is illustrated in Figure 6.6.

In the meshless method, each node is assigned a weight function that is non-zero on a small domain, called the domain of influence. The value of the approximated function at any given point depends on the nodes whose domains of influence contain this point; the approximation is computed from the values of the function at these nodes using a technique known as moving least-squares (MLS) approximation.

The meshless method involves:

– shape functions, constructed by the moving least-squares method;

– the global discrete system, derived from the weak Galerkin formulation;

– the cells of the mesh, used to compute the integrals of K and F.

6.5.2. *Moving Least-Squares Approximation*

The idea of MLS approximation is to construct an approximation locally and then improve it as much as possible. The exact values are not interpolated, but approximated, unlike classical polynomial interpolation.

MLS approximation consists of three components:

– the basis function;

– the weight function assigned to each node;

– a set of coefficients that depend on the position of the node.

Let $u(x)$ be the displacement field on the domain. We want to construct a local approximation in the neighborhood of the point x of the following form:

$$u^h_{local,.x}(x) = \sum_i^m p_i(x)a_i(.x) = p^t(x)a(.x), \qquad [6.3]$$

where m is the number of terms in the polynomial basis, $a(.x)$ is the number of generalized parameters and $p(x)$ is the polynomial basis vector $p^t(x) = \{p_1(x), p_2(x), \ldots, p_m(x)\}$.

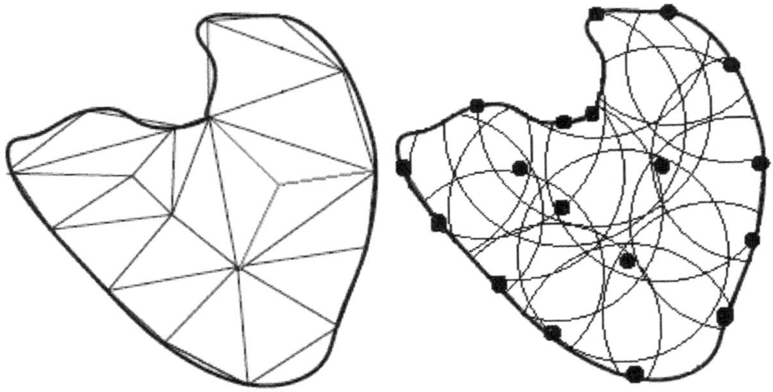

Figure 6.1. *Comparison of the finite element method and the meshless method*

The basis $p(x)$ is constructed using Pascal's triangle. The coefficients a depend on the point x around which the approximation is being constructed.

The global approximation is constructed in such a way as to be equal to this local approximation around any given point. Thus:

$$u^h_{global}(x) = u^h_{local,x}(x) = p^t(x)a(x). \qquad [6.4]$$

The coefficients of $a(x)$ are determined as follows:

– Consider n nodes with positions $x_1, x_2, \ldots, x_n$ at which the values of the displacement $u_1, u_2, \ldots, u_n$ are known. The approximation at the point x_i is then constructed by computing:

$$u^h(x, x_i) = p^t(x_i)a(x), \quad i = 1, 2, \ldots, n. \qquad [6.5]$$

– The norm of the distance between the approximation around x and the known values can be written as:

$$J(x) = \sum_{i=1}^{n} w(x - x_i)[u^h(x, x_i) - u(x_i)]^2 \qquad [6.6]$$

$$= \sum_{i=1}^{n} w(x - x_i)[p^t(x_i)a(x) - u(x_i)]^2. \qquad [6.7]$$

The term $\omega(x - x_i)$ is the weight function, which fulfills a dual purpose: balancing the influence of the point/distance and ensuring that the compatibility condition is satisfied (no dependency between functions).

The coefficients $a(x)$ are computed by minimizing J, which leads to the following linear system:

$$A(x)a(x) = B(x)U. \qquad [6.8]$$

The matrix $A(x)$ has the expression:

$$A(x) = \sum_{i=1}^{n} w_i(x)p(x_i)p^t(x_i) \quad \text{and} \quad w_i(x) = w(x - x_i). \qquad [6.9]$$

Writing:

$$B(x) = [B_1, B_2, \ldots, B_n], \; (B_i = \omega_i(x)p(x_i)), \qquad [6.10]$$

we can now write:

$$a(x) = A^{-1}(x)B(x)U. \qquad [6.11]$$

The MLS approximation of the displacement is:

$$u^h(x) = \sum_{i}^{n} \sum_{j}^{m} p_j(x)(A^{-1}(x)B(x))_{ji}u_i, \qquad [6.12]$$

or, alternatively:

$$u^h(x) = \sum_{i}^{n} \phi_i(x)u_i, \qquad [6.13]$$

where:

$$\phi_i(x) = \sum_{j}^{m} p_j(x)(A^{-1}(x)B(x))_{ji} = p^t A^{-1} B_i. \qquad [6.14]$$

The term $\phi_i(x)$ is the shape function of the MLS approximation at the i-th node, m is the number of terms in the polynomial basis and $n >> m$ is the number of points in the domain of influence.

The approximation of the displacement field can be stated as $u^h(x) = \phi(x)U$. The shape functions ϕ_i are well defined on the domain of approximation if and only if $A(x)$ is invertible at every point x in the domain. The matrix $A(x)$ is a square matrix of dimension equal to the size of the vector p.

6.5.2.1. *Choosing the basis functions*

In general, the basis functions are:

– In 1D: $p^t(x) = 1, x, x^2, \ldots, x^m$,

– In 2D: $p^t(x, y) = 1, x, y, xy, x^2, y^2, \ldots, x^m, y^m$,

– In 3D: $p^t(x, y, z) = 1, x, y, z, xy, yz, zx, x^2, y^2, z^2, \ldots, x^m, y^m, z^m$.

The table below lists the constant, linear and quadratic basis functions in one, two and three dimensions.

	1D	2D	3D
Constant	$[1]$	$[1]$	$[1]$
Linear	$[1, x]$	$[1, x, y]$	$[1, x, y, z]$
Quadratic	$[1, x, x^2]$	$[1, x, y, x^2, y^2, xy]$	$[1, x, y, z, x^2, y^2, z^2, xy, xz, yz]$

Table 6.1. *Basis functions*

We will use the linear basis for the rest of this chapter. This is the most typical choice of basis, for the following two reasons:

– A constant basis is more cost-effective in terms of computation time, but the resulting approximation is not capable of representing linear fields exactly, which is needed for the displacement field in Galerkin methods.

– A quadratic basis would require each point in the domain to be covered by the support of a greater number of functions, and the matrix A that we must invert at each point will be larger than with a linear basis. These additional computations lead to a higher cost/performance ratio in practice.

6.5.2.2. *Choosing the weight functions*

The weight functions are usually chosen to decrease with the distance from their nodes in a bell shape. In one dimension, if s is the normalized distance between the i-th node and an arbitrary point x, then:

$$s = \left| \frac{x_i - x}{d_i} \right|, \qquad [6.15]$$

where d_i is the size of the support of the i-th node.

Common choices of $\omega_i(x)$ include:

– Truncated Gaussian functions:

$$f_1(s) = \begin{cases} e^{-(\frac{s}{\alpha})^2} & \text{if } |s| \le 1 \\ 0 & \text{if } |s| > 1 \end{cases} \qquad [6.16]$$

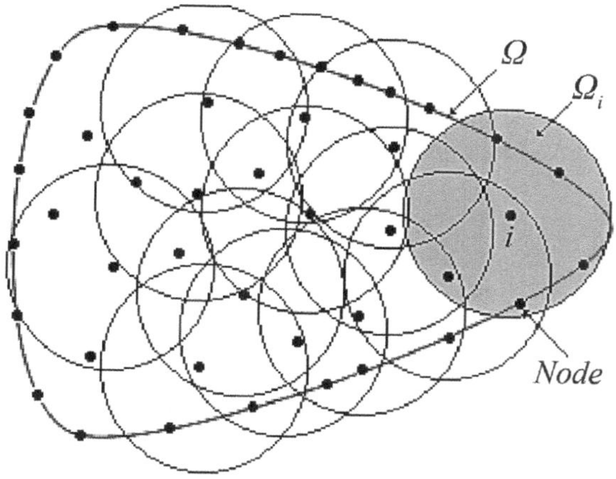

Figure 6.2. *Discretization using a meshless method: nodes, domain of influence (circle)*

This function has the disadvantage of being discontinuous at $s = 1$. In practice, this discontinuity is numerically insignificant when α is sufficiently large.

– Modified Gaussian functions:

$$f_2(s) = \begin{cases} \dfrac{e^{-(\frac{s}{\alpha})^2} - e^{(\frac{1}{\alpha})^2}}{1 - e^{(\frac{1}{\alpha})^2}} & \text{if } |s| \le 1 \\ 0 & \text{if } |s| > 1 \end{cases} \qquad [6.17]$$

This is similar to the truncated Gaussian, but is C^0.

– Cubic splines:

$$f_3(s) = \begin{cases} \frac{2}{3} - 4s^2 + 4s^3 & \text{if } |s| \le \frac{1}{2} \\ \frac{4}{3} - 4s + 4s^2 - \frac{4}{3}^3 & \text{if } \frac{1}{2} < |s| \le 1 \\ 0 & \text{if } |s| > 1 \end{cases} \qquad [6.18]$$

This function is $\mathcal{C}^1$.

– Fourth-order splines:

$$f_4(s) = \begin{cases} 1 - 6s^2 + 8s^3 - 3s^4 & \text{if } |s| \leq 1 \\ 0 & \text{if } |s| > 1 \end{cases} \qquad [6.19]$$

This function is $\mathcal{C}^2$.

In practice, it turns out that the choice of weight function has little influence on the results. Below, we will use fourth-order splines. In two or three dimensions, we must choose one of the above functions with either a circular domain:

$$w_i(x) = f_a\left(\frac{x - x_i}{d_i}\right); \qquad [6.20]$$

or a rectangular domain:

$$w_i(x) = f_a\left(\frac{|x - x_i|}{d_i^x}\right) f_b\left(\frac{|y - y_i|}{d_i^y}\right) \quad \text{(in 2D)}, \qquad [6.21]$$

$$w_i(x) = f_a\left(\frac{|x - x_i|}{d_i^x}\right) f_b\left(\frac{|y - y_i|}{d_i^y}\right) f_c\left(\frac{|z - z_i|}{d_i^z}\right) \quad \text{(in 3D)}. \qquad [6.22]$$

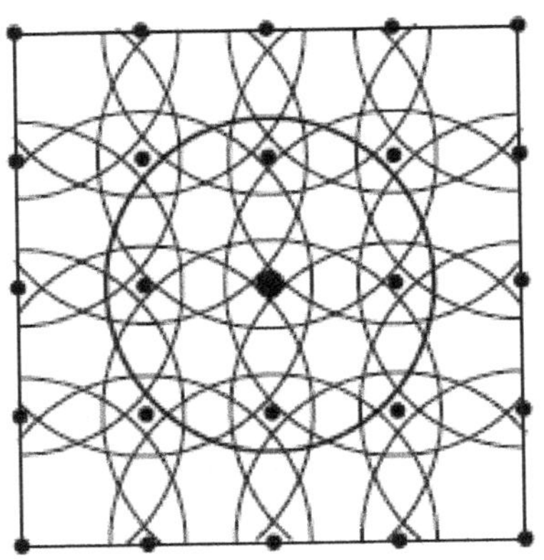

Figure 6.3. *Set of* 5×5 *regularly distributed nodes*

We will use circular domains. The next few figures give an illustration of a two-dimensional problem. Figure 6.3 shows the set of nodes and their support: 5×5 nodes are uniformly distributed over a square domain $[-2, 2] \times [-2, 2]$. The weight functions are fourth-order splines defined on circular domains of radius 1.4. Figure 6.4 shows a plot of the central node on the left – its support can be seen in

Figure 6.3 – and the shape function of this node on the right, using a linear basis. Every point is covered by the support of at least three functions, because the radius was chosen to be sufficiently large; this guarantees that we can compute the shape functions. In Figure 6.4 (two dimensions), we used a set of equidistant nodes with identical weights, but this is not necessary in general – the nodes could alternatively be distributed irregularly. Similarly, the weights can differ from node to node in terms of shape (circular, rectangular, etc.), size (d_i) or type (fourth-order spline, exponential, etc.). The only constraints are that the weights must be positive and there must be sufficiently many non-zero weights at any given point for the approximation to be well defined.

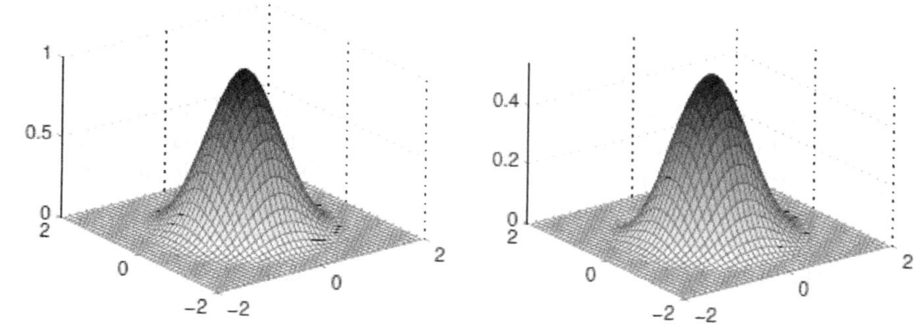

Figure 6.4. *Weight function and shape functions*

Finally, the derivatives of the weight functions can be calculated analytically. For example, consider a fourth-order spline defined on a circular domain. Writing $s = \frac{\|x-x_i\|}{d_i}$, the following relations holds:

$$w_{i,k}(x) = f_4' \frac{x_k - x_{ik}}{s d_i^2} \quad [6.23]$$

$$w_{i,kl}(x) = \left(f''_4 - \frac{f_4'}{s} \right) \frac{(x_k - x_{ik}) - (x_l - x_{il})}{s^2 d_i^4} + f_4' \frac{\delta_{kl}}{s d_i^2}, \quad [6.24]$$

where:

$$f_4'(s) = \begin{cases} -12s + 24s^2 - 12s^3 & \text{if } |s| \leq 1 \\ 0 & \text{if } |s| > 1 \end{cases} \quad [6.25]$$

$$f''_4(s) = \begin{cases} -12 + 48s - 48s^2 & \text{if } |s| \leq 1 \\ 0 & \text{if } |s| > 1 \end{cases} \quad [6.26]$$

To check that this is uniquely defined at $x = x_i$, we observe that $w_{i,k}(x_i) = 0$ and $w_{i,kl}(x_i) = -\dfrac{12\delta_{kl}}{d_i^2}$.

6.5.2.3. *Imposing essential boundary conditions*

The shape functions used by meshless methods are not equal to 1 at their respective nodes:

$$\phi_i(x_j) \neq \delta_{ij} = \begin{cases} 1 & \text{if } i = j \\ 0 & \text{otherwise} \end{cases} \qquad [6.27]$$

where δ_{ij} is the Kronecker delta.

Consequently, the boundary conditions cannot be imposed directly. Various alternative methods have been developed, including the method of Lagrange multipliers [BEL 94], the penalty-based method [ZHU 98], the collocation method [WAG 00] and the coupled meshless/FEM method [HEG 96]. It makes sense to introduce the relevant variational principle before attempting to explain these methods. This variational principle specifies a scalar quantity, the functional Π, defined by the integral formula:

$$\Pi(u) = \int_\Omega F(u, u_{,x}, \ldots) d\Omega + \int_\Omega E(u, u_{,x}, \ldots) d\Gamma, \qquad [6.28]$$

where u is the unknown function, and F and E are differential operators. The solution of this continuous problem is the function u for which Π is stationary under an arbitrary variation δu:

$$\delta\Pi = 0 \quad \forall \delta u. \qquad [6.29]$$

6.5.2.4. *Penalty-based method*

This method adds a penalty term to the weak formulation:

$$\delta\bar{\Pi} = \delta\Pi + \frac{\alpha}{2}\delta\left(\int_{\Gamma_u} \|u - \bar{u}\|^2 d\Gamma\right). \qquad [6.30]$$

This yields the linear system $Ku = f$, where:

$$K_{ij} = \int_\Omega B_i^t C B_j d\Omega - \alpha \int_{\Gamma_u} \phi_i \phi_j d\Gamma \qquad [6.31]$$

$$f_i = \int_{\Gamma_t} \phi_i \bar{t} d\Gamma + \int_\Omega \phi_i b d\Omega - \alpha \int_{\Gamma_u} \phi_i \bar{u} d\Gamma. \qquad [6.32]$$

6.5.2.5. *Method of Lagrange multipliers*

This method is based on the weak formulation. The Lagrange multipliers are used to impose boundary conditions on the displacement. Physically, they can be interpreted as the reaction of the body to its fixation. Consider the general problem of finding the stationary point of the functional Π, subject to the constraint:

$$C(u) = 0 \quad \text{on } \Gamma_u. \tag{6.33}$$

To satisfy these constraints, we construct the following functional:

$$\bar{\Pi}(u, \lambda) = \Pi(u) + \int_{\Gamma_u} \lambda^t C(u) d\Gamma. \tag{6.34}$$

The variation of the new functional is:

$$\delta \bar{P}i = \delta\Pi + \int_{\Gamma_u} \delta\lambda^t C(u) d\Gamma + \int_{\Gamma_u} \lambda^t \delta C(u) d\Gamma. \tag{6.35}$$

To derive the discrete equations, we approximate the Lagrange multipliers as follows:

$$\lambda = \sum_{i=1}^{n} N_i(x)\lambda_i. \tag{6.36}$$

We will explain this method in more detail below.

By comparison with the Lagrange multiplier method, the greatest advantage of the penalty-based method is that it does not require any additional unknowns. However, it can introduce discrepancies if the penalty parameter is poorly designed.

6.5.2.6. *Integration*

To compute the stiffness matrix K and the load vector f in the matrix formulation $KU = f$, we need to evaluate some of the integral terms. We need to know the value of the integrals along certain contours to find the contribution of the surface forces to the force vector and, for some problems, the contribution of the essential boundary conditions to the stiffness matrix, depending on the method. Several approaches have been used to carry this out. The following two are among the most widely used:

– Integration on an implicit mesh. The simplest method is to apply the classical integration formula to a mesh (Figure 6.5). We can distribute integration points over each element/cell, like in the FEM, according to one of the two possible techniques: the first is to construct a mesh whose elements are connected by the nodes of the MLS approximation (Figure 6.5, left), and the second technique is to uniformly divide the problem into cells that cover the component and which are independent of the approximation nodes. Gaussian integration is then performed on each cell, assigning zero weight to any Gaussian points outside of the component (Figure 6.5, right). Constructing the mesh on the point cloud (left of 6.5) preserves the external boundaries of the domain, which eliminates integration errors. However, the fact that the support of the shape functions does not coincide perfectly with the elements/cells represents a source of errors.

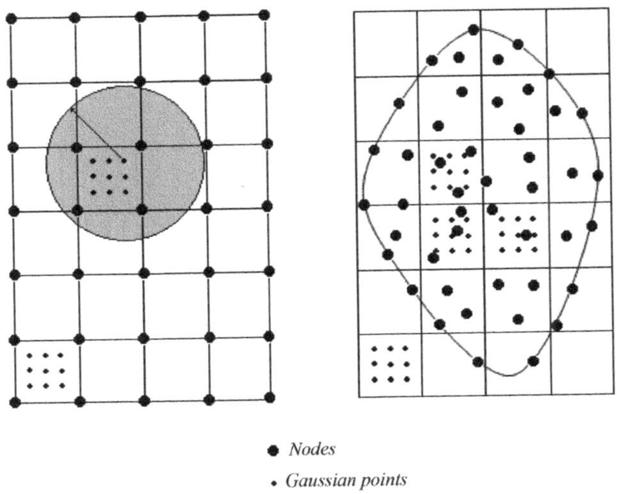

Figure 6.5. *Implicit meshes*

Even though the method is no longer strictly meshless, the problems encountered by the FEM do not necessarily resurface. The implicit mesh is not required to follow the interior boundaries of the computation domain and is not subject to the same constraints: the shape functions are never degenerate, because their support is not bound to the mesh. Hence, we do not need to keep remeshing the domain, and the mesh is less computationally expensive. In summary, the key difference compared with the FEM is that the mesh is only used for numerical integration, and is not part of the approximation scheme. The greatest challenge of the meshless approach lies in choosing the integration points. There must be a suitable number of integration points relative to the number of discretization nodes: if there are too few integration points in the zone of influence of a node, then the precision of the method suffers, and artificial modes can be introduced. By contrast, if there are too many integration points, the computations become excessively heavy (the shape functions can sometimes be expensive to evaluate, and need to be evaluated once per integration point). A poor distribution of integration points can negatively affect the conditioning of the matrix that needs to be inverted.

– Direct nodal integration. If we want to avoid an implicit mesh, then we need a way of formulating the integrals in terms of the nodal values u_i at the discretization points x_i only, without introducing any additional points. Any such method is said to be a direct nodal integration technique. These integration schemes are typically unstable and give rise to parasitic modes, because the gradient of ϕ_i vanishes at x_i in general. A stabilized nodal integration technique, called stabilized conformal nodal integration, has been recently proposed by Chen *et al.* [CHE 01].

We will now explore the first approach further (see Figure 6.6).

6.6. Application of the meshless method to elasticity

6.6.1. *Formulation of static linear elasticity*

Consider the following two-dimensional linear elasticity problem. The domain Ω is enclosed by its boundary Γ:

$$L^t\sigma + b = 0 \quad \text{on } \Omega, \qquad [6.37]$$

subject to the boundary conditions:

$$\begin{cases} \sigma.n = \bar{t} & \text{on } \Gamma_t \\ u = \bar{u} & \text{on } \Gamma_u \end{cases} \qquad [6.38]$$

where u, σ, b and n, respectively, denote the displacement field, the stress tensor, the force per unit volume and the normal unit vector pointing outward from the boundary Γ. The terms $\bar{u}$ and $\bar{t}$, respectively, denote the displacements and tractions imposed on the displacement and traction boundaries Γ_u and Γ_t. The matrix L has the expression:

$$L = \begin{bmatrix} \frac{\partial}{\partial x} & 0 \\ 0 & \frac{\partial}{\partial y} \\ \frac{\partial}{\partial y} & \frac{\partial}{\partial x} \end{bmatrix}.$$

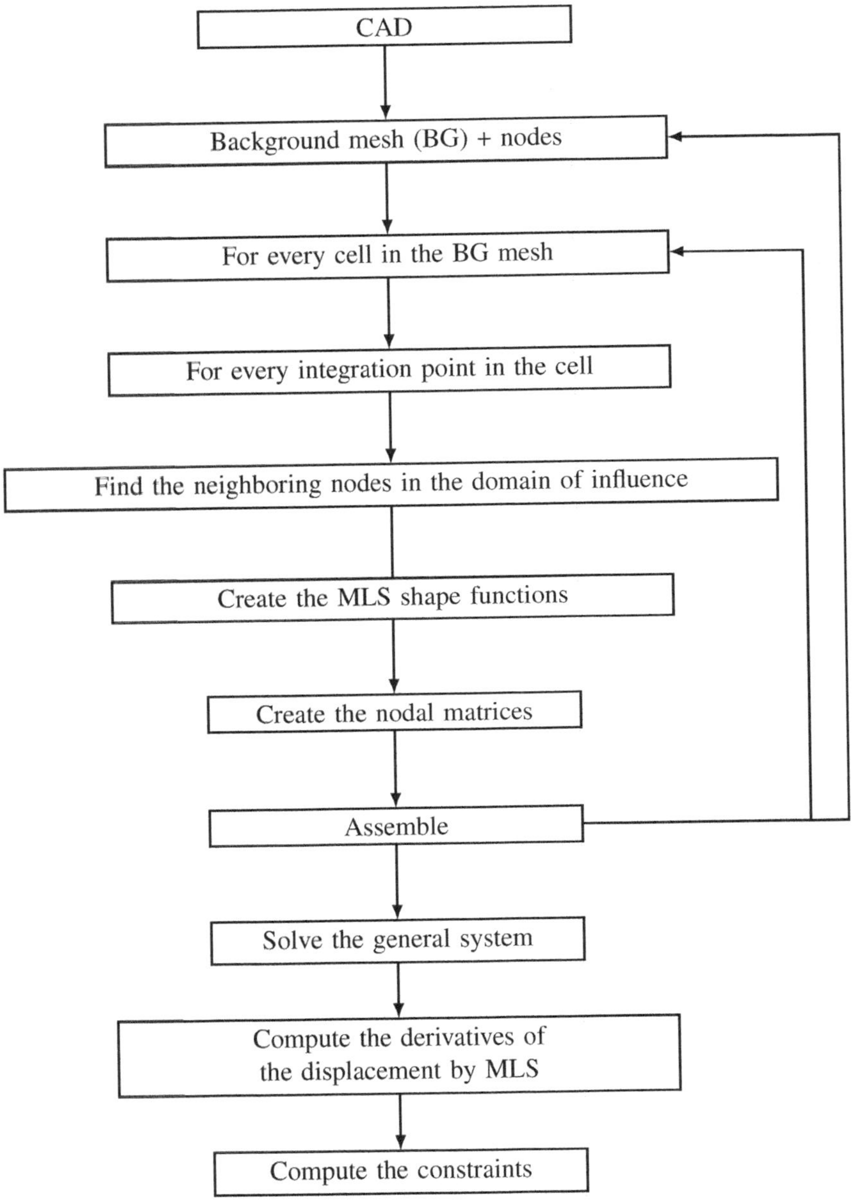

Figure 6.6. *General algorithm of the EFG method*

The variational formulation (principle of virtual work) associated with this elastostatic problem can be stated as follows:

$$\int_{\Omega} \delta(Lu)^t(CLu)d\Omega - \int_{\Omega} \delta u^t b d d\Omega - \int_{\Gamma_t} \delta u^t \bar{t} d\Gamma = 0. \qquad [6.39]$$

The MLS approximation of the $u_h(x)$ component of the displacement is given by equation [6.13]. Similarly, the MLS approximation of the $v_h(x)$ component is:

$$v_h(x) = \sum_i^n \phi_i(x) v_i. \qquad [6.40]$$

Grouping these two components together gives:

$$U_h = \sum_i^n \begin{bmatrix} \phi_i & 0 \\ 0 & \phi_i \end{bmatrix} \begin{pmatrix} u_i \\ v_i \end{pmatrix} = \sum_i^n \Phi_i U_i, \qquad [6.41]$$

where Φ_i is the matrix of shape functions. Thus:

$$LU^h = \sum_i^n L\Phi_i U_i = \sum_i^n \begin{bmatrix} \phi_{i,x} & 0 \\ 0 & \phi_{i,y} \\ \phi_{i,y} & \phi_{i,y} \end{bmatrix} U_i = \sum_i^n B_i u_i, \qquad [6.42]$$

where B_i is the matrix of deformations at the i-th node.

6.6.2. *Imposing essential boundary conditions*

Essential boundary conditions can be imposed with Lagrange multipliers. The variational formulation is:

$$\int_{\Omega} \delta(Lu)^t(CLu)d\Omega - \int_{\Omega} \delta u^t b d d\Omega - \int_{\Gamma_t} \delta u^t \bar{t} d\Gamma$$
$$- \int_{\Gamma_u} \delta\lambda^t (u - \bar{u}) d\Gamma - \int_{\Gamma_u} \delta u^t \lambda d\Gamma = 0, \quad [6.43]$$

where λ is the vector of Lagrange multipliers.

These Lagrange multipliers are unknown, so we approximate them along the contour:

$$\lambda(x) = \sum_i^{n_\lambda} N_i(s)\lambda_i \quad x \in \Gamma_u, \qquad [6.44]$$

where n_λ is the number of nodes used for the interpolation, s is the arclength, λ_i is the Lagrange multiplier at the i-th node and N_i is the corresponding Lagrange-type shape function (e.g. $N_0(s) = \frac{s-s_1}{s_0-s_1}$, $N_1(s) = \frac{s-s_0}{s_1-s_0}$).

Varying the Lagrange multiplier gives:

$$\delta\lambda(x) = \sum_i^{n_\lambda} N_i(s)\delta_i \quad x \in \Gamma_u. \qquad [6.45]$$

The vector of multipliers is therefore:

$$\lambda = \sum_{i=1}^{n_\lambda} \underbrace{\begin{bmatrix} \mathrm{N}_i & 0 \\ 0 & \mathrm{N}_i \end{bmatrix}}_{N_i} \underbrace{\begin{Bmatrix} \lambda_{u_i} \\ \lambda_{v_i} \end{Bmatrix}}_{\lambda_i} = \sum_{i=1}^{n_\lambda} N_i \lambda_i. \qquad [6.46]$$

Substituting these approximations into the variational formulation (equation [6.43]) gives:

$$\int_\Omega \delta(\sum_i^n B_i u_i)^t (C \sum_i^n B_j u_j) d\Omega - \int_\Omega \delta(\sum_i^n \phi_i u_i)^t b d\Omega - \int_\Gamma \delta(\sum_i^n \phi_i u_i)^t \bar{t} d\Gamma$$

$$- \int_{\Gamma_u} \delta\lambda^t ((\sum_i^n \phi_i u_i)^t - \bar{u}) d\Gamma - \int_{\Gamma_u} \delta\lambda^t (\sum_i^n \phi_i u_i)^t \lambda d\Gamma = 0. \qquad [6.47]$$

The first term of equation [6.47] is:

$$\int_\Omega \delta(\sum_i^n B_i u_i)^t (C \sum_i^n B_j u_j) d\Omega = \int_\Omega \delta(\sum_i^n u_i^T B_i^t)(C \sum_i^n B_j u_j) d\Omega. \qquad [6.48]$$

The summation, integration and variation operators are all linear:

$$\int_\Omega \delta(\sum_i^n u_i^t B_i^t)(C \sum_i^n B_j u_j) d\Omega = \sum_i^n \sum_j^n \delta u_i^t \underbrace{\int_\Omega B_i^t C B_j d\Omega u_j}_{K_{ij}}$$

$$= \sum_i^n \sum_j^n \delta u_i^t K_{ij} u_j, \qquad [6.49]$$

where K_{ij} is the 2×2 nodal stiffness matrix and n_t is the total number of nodes.

The sum above represents an assembly operation:

$$\begin{aligned}\sum_i^{n_t}\sum_j^{n_t}\delta u_i^t K_{ij}u_j &= \delta u_1^t K_{11}u_1 + \delta u_1^t K_{12}u_2 + \cdots + \delta u_1^t K_{1n_t}u_{n_t}\\ &\quad +\delta u_2^t K_{21}u_1 + \delta u_2^t K_{22}u_2 + \cdots + \delta u_2^t K_{2n_t}u_{n_t}\\ &\qquad\vdots \qquad\qquad \vdots \qquad\qquad \vdots\\ &\quad +\delta u_{n_t}^t K_{n_{t1}}u_1 + \delta u_{n_t}^t K_{n_{t2}}u_2 + \cdots + \delta u_{n_t}^t K_{n_t n_t}u_{n_t}\\ &= \delta U^t K U, \end{aligned} \qquad [6.50]$$

where K is the $2n_t \times 2n_t$ global stiffness matrix:

$$K = \begin{bmatrix} K_{11} & K_{12} & \cdots & K_{1n_t}\\ K_{21} & K_{22} & \cdots & K_{2n_t}\\ \vdots & \vdots & \ddots & \vdots\\ K_{n_t1} & K_{12} & \cdots & K_{n_t n_t}\end{bmatrix} \qquad [6.51]$$

and U is the $2n_t \times 1$ global displacement vector:

$$U = \begin{Bmatrix} u_1\\ u_2\\ \vdots\\ u_{n_t}\end{Bmatrix}. \qquad [6.52]$$

The vector of external forces can be derived from the second term of the variational formulation (equation [6.43]):

$$\begin{aligned}\int_\Omega \delta u^t b d\Omega &= \int_\Omega \delta(\sum_i^n \phi_i u_i)^t b d\Omega\\ &= \sum_i^n \delta u_i^t \underbrace{\int_\Omega \phi_i^t b d\Omega}_{f_i}\\ &= \sum_i^n \delta u_i^t f_i, \end{aligned} \qquad [6.53]$$

where f_i is the vector of nodal forces.

This sum can be rewritten as:

$$\sum_{i}^{n} \delta u_i^t f_i = \delta u_1^t f_1 + \delta u_2^t f_2 + \cdots + \delta u_n^t f_n = \delta U^t F. \qquad [6.54]$$

The third term of equation [6.43] completes the vector of external forces, following the same approach as above, with:

$$f_i = \int_{\Gamma_t} \phi_i^t \bar{t} d\Gamma. \qquad [6.55]$$

The fourth term of the variational formulation (equation [6.43]) is:

$$\begin{aligned}
\int_{\Gamma_u} \delta\lambda^t (u - \bar{u}) d\Gamma &= \int_{\Gamma_u} \delta\lambda^t ((\sum_{i}^{n} \phi_i u_i) - \bar{u}) d\Gamma \\
&= \int_{\Gamma_u} \delta(\sum_{i}^{n_\lambda} N_i \lambda_i)^t \sum_{j}^{n} \phi_j u_j d\Gamma - \int_{\Gamma_u} \delta(\sum_{i}^{n_\lambda} N_i \lambda_i)^t \bar{u} d\Gamma \\
&= \sum_{i}^{n_\lambda} \sum_{j}^{n_t} \delta\lambda_i^t \underbrace{\int_{\Gamma_u} N_i^t \phi_j d\Gamma u_j}_{-G_{ij}^t} - \sum_{i}^{n\lambda} \delta\lambda_i^t \underbrace{\int_{\Gamma_u} N_i^t \bar{u} d\Gamma}_{-q_i} \qquad [6.56] \\
&= \sum_{i}^{n_{\lambda t}} \sum_{j}^{n_t} \delta\lambda_i^t G_{ij}^t u_j + \sum_{j}^{n_{\lambda t}} \delta\lambda_i^t q_i \\
&= -\delta\lambda^t G^t U + \delta\lambda^t q.
\end{aligned}$$

The vector q, of size $2n_t$, is the vector of imposed displacements. The fifth term in the variational formulation (equation [6.43]) is:

$$\begin{aligned}
\int_{\Gamma_u} \delta u^t \lambda d\Gamma &= \int_{\Gamma_u} \delta(\sum_{i}^{n} \phi_i u_i)^t \lambda d\Gamma \\
&= \int_{\Gamma_u} \delta(\sum_{i}^{n} \phi_i u_i)^t (\sum_{i}^{n_\lambda} N_j \lambda_j) d\Gamma \\
&= \sum_{i}^{n} \sum_{j}^{n_\lambda} \delta u_i^t \underbrace{\int_{\Gamma_u} \phi_j N_i^t d\Gamma \lambda_j}_{-G_{ij}} \qquad [6.57]
\end{aligned}$$

$$= -\sum_{i}^{n}\sum_{j}^{n_{\lambda t}} \delta u_i^t G_{ij}\lambda_j$$

$$= -\delta U^t G\lambda,$$

where n_λ is the number of nodes on the boundary affected by the boundary conditions of the i-th node, $n_{\lambda t}$ is the number of total nodes on the contour Γ_u, G_{ij} is the 2×2 nodal matrix and G is the global $2n_t \times 2n_t$ matrix.

After replacing all of the matrix parameters in the variational formula, we find:

$$\underbrace{\int_\Omega \delta(\sum_i^n u_i^t B_i^t)(C\sum_i^n B_j u_j)d\Omega}_{\delta U^t KU} - \underbrace{\int_\Omega \delta(\sum_i^n \phi_i u_i)^t b d\Omega - \int_\Gamma \delta(\sum_i^n \phi_i u_i)^t \bar{t} d\Gamma}_{\delta U^t F}$$

$$-\underbrace{\int_{\Gamma_u} \delta\lambda^t((\sum_i^n \phi_i u_i)^t - \bar{u})d\Gamma}_{\delta\lambda^t(G^tU-q)} - \underbrace{\int_{\Gamma_u} \delta\lambda^t(\sum_i^n \phi_i u_i)^t \lambda d\Gamma}_{\delta U^t G\lambda} = 0. \qquad [6.58]$$

This can be written as:

$$\delta U^t(KU + G\lambda - F) + \delta\lambda^t(G^tU - q) = 0. \qquad [6.59]$$

Hence:

$$\begin{cases} KU + G\lambda - F = 0 \\ G^tU - q = 0 \end{cases} \qquad [6.60]$$

Equivalently:

$$\begin{bmatrix} K & G \\ G^t & 0 \end{bmatrix} \begin{Bmatrix} U \\ \lambda \end{Bmatrix} = \begin{Bmatrix} F \\ q \end{Bmatrix}, \qquad [6.61]$$

where:

$$K_{ij} = \int_\Omega B_i^t C B_j d\Omega \qquad [6.62]$$

$$G_{ij} = -\int_{\Gamma_u} \phi_i^t N_j d\Gamma \qquad [6.63]$$

$$F_{ij} = \int_\Omega \phi_i^t b d\Omega + \int_{\Gamma_t} \phi_i^t \bar{t} d\Gamma \qquad [6.64]$$

$$q_i = -\int_{\Gamma_u} N_i^t \bar{u} d\Gamma \qquad [6.65]$$

$$B_i = L\phi_i = \begin{bmatrix} \phi_{i,x} & 0 \\ 0 & \phi_{i,x} \\ \phi_{i,y} & \phi_{i,y} \end{bmatrix} \qquad [6.66]$$

$$N = \begin{bmatrix} N_i & 0 \\ 0 & N_i \end{bmatrix} \qquad [6.67]$$

$$\phi = \begin{bmatrix} \phi_i & 0 \\ 0 & \phi_i \end{bmatrix}. \qquad [6.68]$$

In equation [6.67], N_i are the conventional shape functions from the finite element method, which are used to approximate the Lagrange multipliers under the boundary conditions.

6.7. Numerical examples

6.7.1. *Fixed-free beam*

In this example, we use linear basis functions and cubic spline weight functions for the MLS approximation. Each cell is integrated using 4×4 Gaussian points.

Consider a fixed-free beam subject to a concentrated force [PHU 07] (Figure 6.7). The beam is assumed to be in a plane stress state.

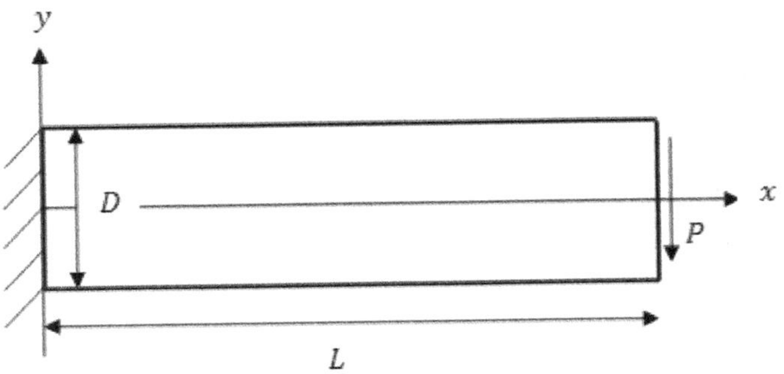

Figure 6.7. *Fixed-free beam subject to a concentrated force*

The geometric parameters of the computation are as follows: $E = 3 \times 10^6$, $\nu = 0.3$, $P = 1,000$, $L = 100$ and $D = 10$. We will use 18×7 nodes, distributed as shown in Figure 6.8. After using the EFG method to compute the displacements, stresses and strains of the problem, the results are shown in Figure 6.9.

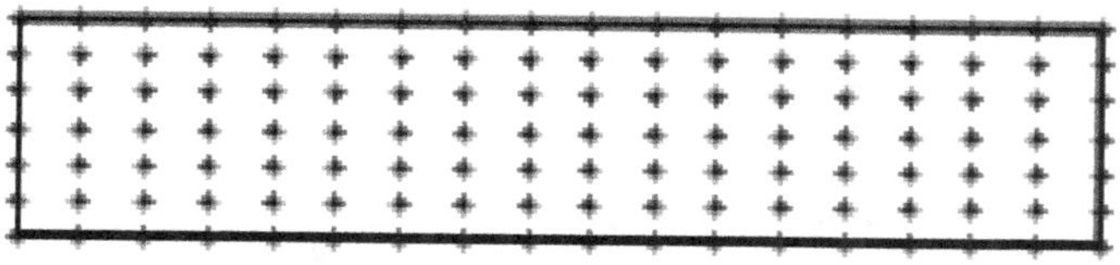

Figure 6.8. *Distribution of the nodes*

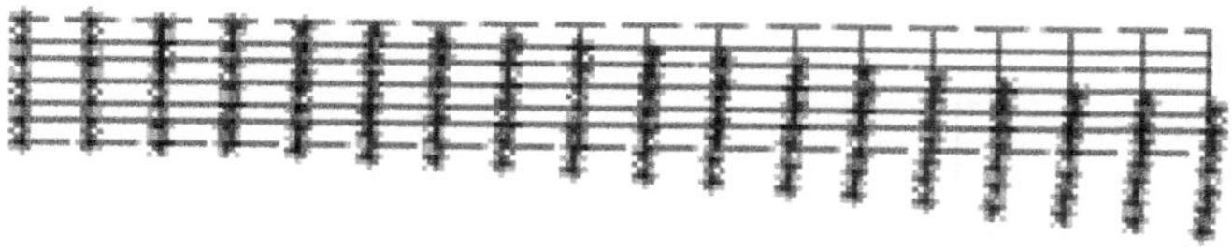

Figure 6.9. *Visualization of the deformation*

6.7.2. *Compressed block*

Consider a metal block under compression (Figure 6.10). By the symmetry of the model, we only need to consider half of the block (Figure 6.11). The geometric parameters are as follows: $E = 2.1 \times 10^5$ and $\nu = 0.3$.

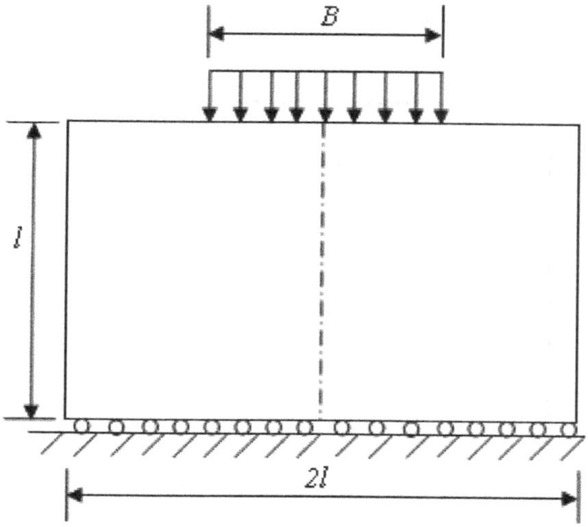

Figure 6.10. *Metal block*

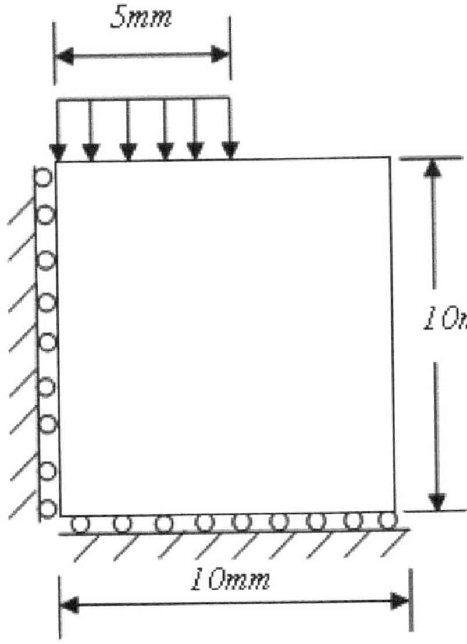

Figure 6.11. *Geometry and boundary conditions*

Figure 6.12 shows the distribution of the 289 nodes over the domain. The results are computed with the assumption of a vertical displacement of −0.2 mm. The metal block is assumed to be in a plane strain regime (see Figure 6.13).

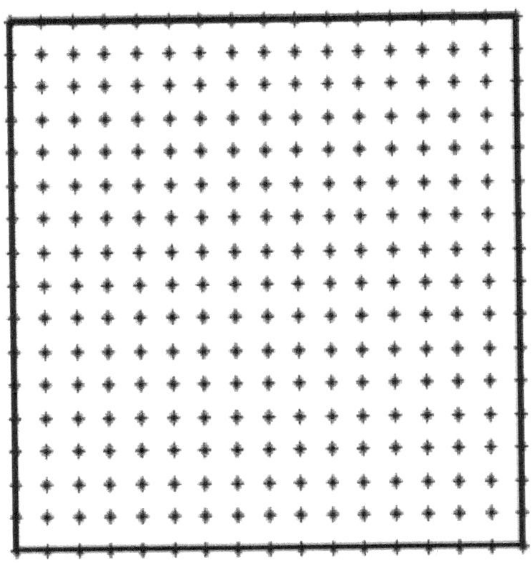

Figure 6.12. *Distribution of nodes*

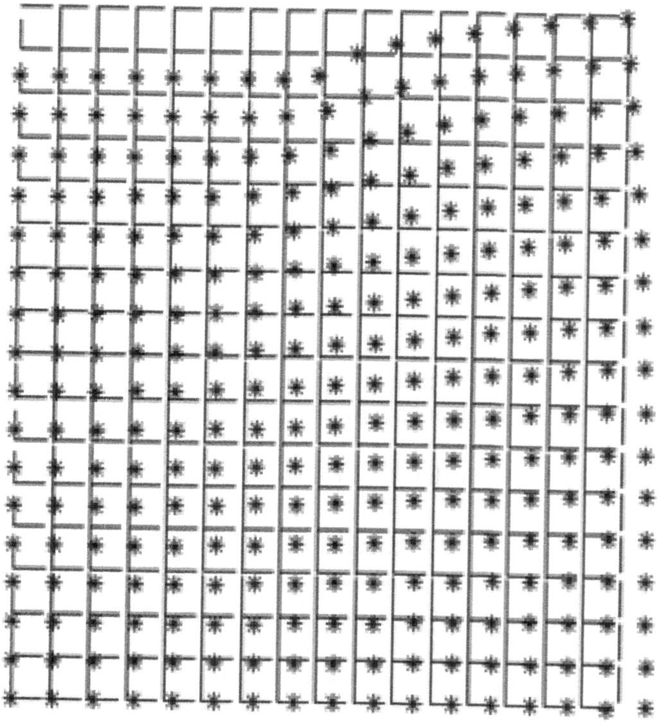

Figure 6.13. *Visualization of the deformation*

6.8. Using *Matlab*

A comprehensive library of example *Matlab* scripts for this method can be found in [FAS 07]. For instance, consider the Poisson problem:

$$\Delta u(x,y) = 13e^{(-2x+3y)} \quad \text{on } \Omega \qquad [6.69]$$

$$u(x,y) = e^{(-2x+3y)} \quad \text{on } \partial\Omega \qquad [6.70]$$

where the domain Ω is the unit disk.

The exact solution of this problem is:

$$u(x,y) = e^{(-2x+3y)}. \qquad [6.71]$$

By using a particular solutions approach to approximate the Laplacian with the following multiquadratic basis functions (MQ):

$$\phi(r) = \sqrt{r^2 + c^2}; \tag{6.72}$$

and the following particular solutions:

$$\phi(r) = \frac{1}{9}(4c^2 + r^2)\sqrt{r^2 + c^2} - \frac{c^3}{3}\ln(c + \sqrt{r^2 + c^2}); \tag{6.73}$$

we will solve the Poisson problem on 300 randomly distributed points in the domain and 40 boundary points. We will then test the result against various parameters c and find the maximum error at another 200 random points in the domain.

The *Matlab* script is listed below [MAT 10]:

```
%-----------------------------------------------
% This script is based on the MAPS method by
%     C.S. Chen, C.M. Fan and P.H. Wen.
% c: Optimal shape parameter.
%------------------------------------------------

n_in = 300;           % number of interior points
n_b  = 40;           % number of boundary points
n = n_in + n_b;

% Random points in the disk
    rad = rand(n_in,1);
    th = 2*pi * rand(n_in,1);
    x1 = sqrt(rad) .* cos(th);
    y1 = sqrt(rad) .* sin(th);

% Points on the circle
    pi2 = 2 * pi;
    st = pi2/n_b;
    ro = 0 : st : pi2-st;
    x2 = cos(ro');
    y2 = sin(ro');

% Radial basis functions and their solutions
    R = [x1, y1; x2, y2];      % All points in the matrix
    r = squareform(  pdist(R,'euclidean')  );
```

```
    r2 = r.^2;
    c2 = c^2;
    c3 = c^3/3;

    fi1 = sqrt(r2(1:n_in,:) + c2);  % MQ: radial basis functions
    fi2 = sqrt(r2(n_in+1:n,:) + c2);
    FI=(4*c2 + r2(n_in+1:n,:)).*fi2/9 - c3*log(c + fi2); %
         corresponding particular solutions

% Exact solution on random points selected above
    u_exact = exp(-2*R(:,1) + 3*R(:,2));

% System of equations
    A = [fi1; FI];
    a = zeros(n,1);

    b = exp(-2*R(:,1) + 3*R(:,2));
    b(1:n_in) = 13 * b(1:n_in);

% Singular value decomposition (SVD)
    [U,Sing,V] = svd(A);

    % Test
    e = log10(max(max(Sing)))-16 + 16/5;
    epsilon = 10^(e);

    m = length(A);
    k = 0;
    for i = 1:m
        if Sing(i,i) < epsilon
            Sing(i,i)=0;
            k = k+1;
        else Sing(i,i) = 1/Sing(i,i);
        end
    end
    k;
    A = V * Sing * U';
    a = A * b;  % Coefficients - solution of SVD.

% Results
  % Random points selected above
    fi = [fi1;fi2];
```

```
    FI_comp = (4*c2 + r2) .* fi / 9 - c3 * log(c + fi);
    u_comp = sum((FI_comp*diag(a))') ;
    RMSE = sqrt(sum( (u_comp(1:n_in)'-u_exact(1:n_in)).^2 )/n_in);

  % New random points
    n_t = 200;
    rad = rand(n_t,1);
    th = 2*pi * rand(n_t,1);
    x3 = sqrt(rad) .* cos(th);
    y3 = sqrt(rad) .* sin(th);
    R_t = [x3, y3; R];      % All points in the matrix;
    r_t = squareform( pdist(R_t,'euclidean') );
    r_t2 = r_t(n_t+1:end,1:n_t).^2;
    fi_t = sqrt(r_t2 + c2);     % MQ: radial basis functions
    FI_t = (4*c2 + r_t2) .* fi_t / 9 - c3* log(c + fi_t); %
            corresponding particular solutions
    u_t = sum((diag(a)*FI_t));
    u_t_exact = exp(-2*x3 + 3*y3);
    RMSE_t = sqrt(sum( (u_t'-u_t_exact).^2 )/n_t);

% Plot the result
  % Plot u
    x_int = linspace(min(R(:,1)),max(R(:,1)),500);
    y_int = (linspace(min(R(:,2)),max(R(:,2)),500))';
    [X,Y,Z] = griddata(R(:,1),R(:,2),u_comp',x_int,y_int,'linear');
    mesh(X,Y,Z);
    xlabel('x');
    ylabel('y');
    zlabel('u(x,y)');
```

The mesh is shown in Figure 6.14.

The formula of the RMSE is given by equation [6.74]. The error is less than 10^{-4} at every point. The solution of the equation is shown in Figure 6.16.

$$\text{RMSE} = \sqrt{\frac{1}{m}\sum_{j=1}^{m}(\hat{u}_j - u_j)^2}. \qquad [6.74]$$

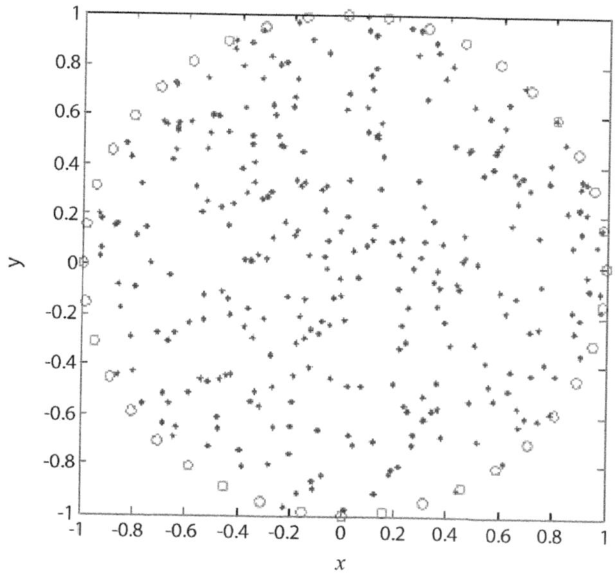

Figure 6.14. *The n_i random points in the domain and n_b points on the boundary. For a color version of this figure, see www.iste.co.uk/radi/advanced2.zip*

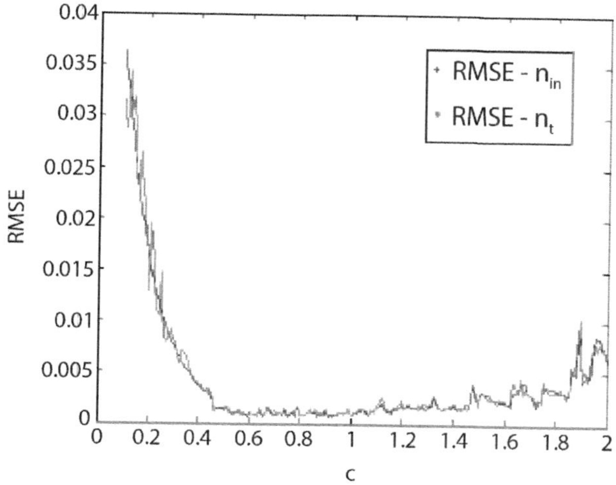

Figure 6.15. *RMSE as a function of the shape parameter c for two sets of values. For a color version of this figure, see www.iste.co.uk/radi/advanced2.zip*

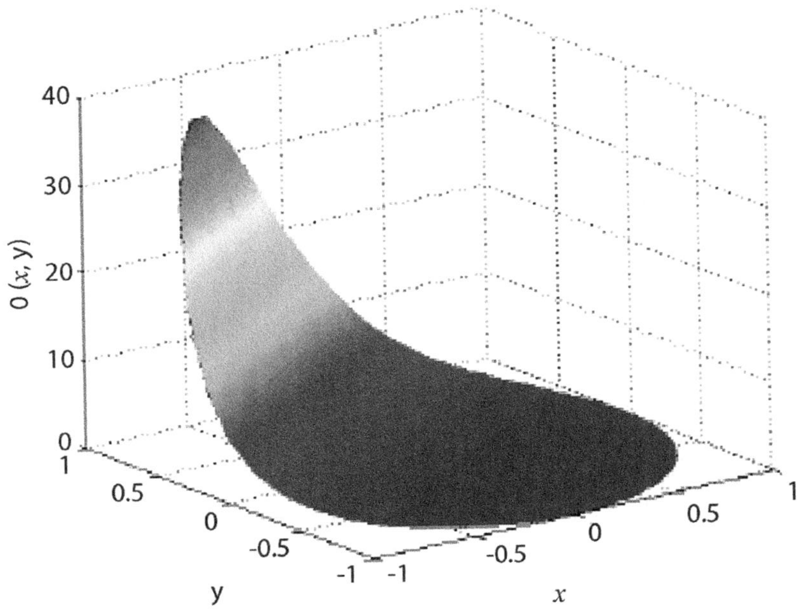

Figure 6.16. *Profile of the solution of the problem. For a color version of this figure, see www.iste.co.uk/radi/advanced2.zip*

PART 3

Appendices

Appendix 1

Introduction to *Matlab*

A1.1. Introduction

Matlab stands for matrix laboratory. Originally written in Fortran by Cleve Moler, *Matlab* was designed to make it easier to access the matrix software developed by the LINPACK and EISPACK projects. The modern version of *Matlab* is written in C, published by MathWorks Inc. and is available in both professional and student versions across multiple platforms.

Matlab is a powerful, comprehensive and easy-to-use environment for scientific computation. It gives engineers, researchers and scientists an interactive system that integrates numerical calculations and visualizations.

Matlab uses its own intuitive and natural programming language, which offers spectacular CPU performance improvements over other languages such as C, TurboPascal and Fortran. *Matlab* allows users to dynamically include links to programs in C or Fortran, exchange data with other software applications or use *Matlab* itself as the engine for analysis and visualization.

Matlab also offers specialized tools for certain fields, known as ToolBoxes, which are regarded as one of *Matlab*'s most attractive features for most users. These ToolBoxes are collections of functions that extend the *Matlab* environment to enable it to solve certain types of problems. They cover a wide range of topics, including signal processing, automation, neural networks, structural computations and statistics.

Matlab allows users to work interactively in either command mode or programming mode; graphical visualizations can be generated in either case. Widely viewed as one of the best programming languages (alongside others such as C or Fortran), *Matlab* offers the following specific advantages relative to its competitors:

– easy programming;

– continuity between integer, real and complex values;

– an extensive range of numbers and precisions;

– a very comprehensive mathematical library;

– graphical tools, including graphical interface tools and utilities;

– the ability to add links to other classical programming languages (e.g. C or Fortran).

The graphical interface allows scientific or even artistic representations to be generated from their mathematical expressions. The figures generated by *Matlab* are simple and eye-catching, and an impressive array of features is available to enhance them.

A1.2. Starting up *Matlab*

To launch *Matlab*:

– in Windows, click on Start, then Programs, then *Matlab*;

– for other operating systems, refer to the user manual.

The *Matlab* prompt ">>" should then be displayed. This is where users can enter commands (see Figure A1.1).

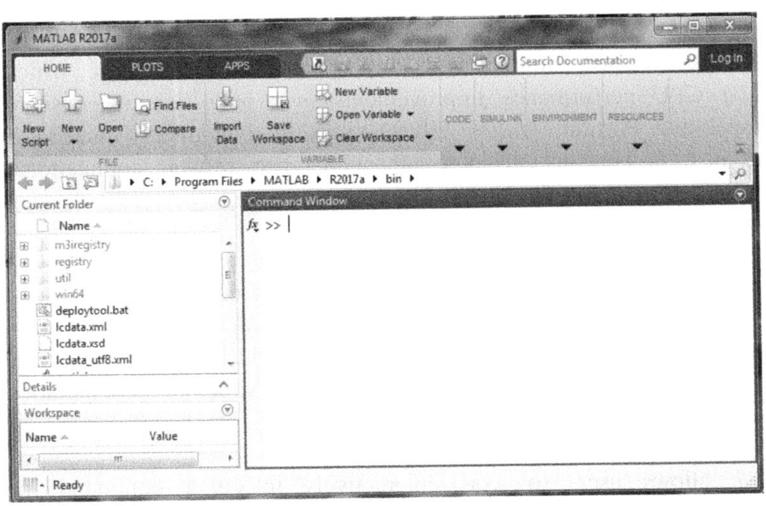

Figure A1.1. *Matlab window*

The *quit* function allows the user to exit *Matlab*:

```
>>quit
```

The "help" command gives information about specific problems. For example:

```
>> help cos

COS Cosine.
COS(X) is the cosine of the elements of X.
```

The standard arithmetic operations are supported:

+ : addition; - : subtraction; / : division; * : multiplication; ^ : exponentiation; pi=π. For example:

```
>>x=2

x =

     2
>>P=(4*x^2-2*x+3)/(x^3+1)

P =

     1.6667
```

Command mode allows us to perform computations in *Matlab*.

For example, suppose that we wish to calculate the following volume: $V = \frac{4}{3}\pi R^3$, where $R = 4$ cm. We can carry this out as follows:

```
>>R=4

R =

     4
>>V=4/3*pi*R^3

V =

     268.0826
```

A1.3. Mathematical functions

The trigonometric functions are:

sin(x), cos(x), tan(x), asin(x), acos(x), atan(x), atan2(x,y), sinh(x), cosh(x), tanh(x), asinh(x), acosh(x), atanh(x).

The other (elementary) mathematical functions available in *Matlab* are:

abs(x)	Absolute value of x
angle(x)	If x is a positive real number, then angle(x) = 0; if x has positive imaginary part, then angle(x) = pi/2
sqrt(x)	Square root of x
real(x)	Real part of the complex value x
imag(x)	Imaginary part of the complex value x
conj(x)	Complex conjugate of x
round(x)	Round x to the nearest integer
fix(x)	Round x toward zero
floor(x)	Round x to the integer below
ceil(x)	Round x to the integer above
sign(x)	=+1 if x>0; =-1 if x<0
rem(x,y)	The remainder of the division: = x-y*fix(x/y)
exp(x)	Exponential (base e)
log(x)	Logarithm (base e)
log10(x)	Logarithm (base 10)

A1.4. Operators and programming with *Matlab*

Most programming scripts make frequent use of "if" statements. Every "if" statement must be followed by the "end" keyword:

```
>>V=268.0826

V =
    268.0826
>> R = 4

R =
    4
>>if V>150, surface=pi*R^2, end

surface =
    50.2655
```

The "not" operator is written (or symbolically represented) by "~=":

```
>>R=4
```

```
R =
     4
>>if R ~=2, V=4/3*pi*R^3, end

V=
     268.0826
```

The "equals" operator (==) in an "if" statement is written (or symbolically represented) by "==":

```
>>R=4
R =
     4
>>if R==4, V=4/3*pi*R^3; end
```

The "or" operator is written (or symbolically represented) by "|". For example, the test "if $R = 4$ or $m = 1$" can be written as:

```
>>if R==4 | m==1, V=4/3*pi*R^3; end
```

Other operators available in *Matlab* include:

>	greater than	<	less than
>=	greater than or equal to	<=	less than or equal to
^	power (exponent)	*	product
xor	exclusive OR (XOR)	/	division
Error	displays the message: "error"		

For example, "if $g > 2$ or $g < 0$, then $a = 4$" can be programmed as follows:

```
>>if g>2 |g<0,
a=4;
end
```

As another example, "if a > 3 and c < 0, then b = 15" can be implemented with the code:

```
>>if a>3 \& c<0,
b=15;
end
```

The operators "&" and "|" can be chained together:

```
>>if ((a==2 | b==3)\&(c<5)),
g=1;
end
```

The operator "if ... elseif ... else ... end" can be used as follows:

```
>>R=2, if R>3, b=1 ;
elseif R==3, b=2;
else b=0;
end
```

The keyword "elseif" can be repeated as many times as necessary within the same program.

The "for/end" and "while/end" operators can, for example, be used as follows:

```
>>for R=1 :5,
V=4/3*pi*R^3;
disp([R,V]);
end
```

In this example, R ranges from 1 to 5 and the command "disp([R,V])" returns the matrix [R=1 :5,V (V(1) :V(5)].

The "length" command can also be called, returning the size of a variable. In the above example, length(R)=5; (R=1 :5) and length(R)-1=4 (4 intervals separated by increments of 1).

```
>>while R<5, R=R+1 ; V=4/3*pi*R^3; disp([R,V]); end
```

The "while" command continues to execute the same instruction until its logical test no longer returns true.

The next example demonstrates how to use an automatic increment in a "for" loop:

```
>>for R=5 :-1 :1, V=4/3*pi*R^3; disp([R,V]); end
```

Here, the increment is decreasing (=-1). "For" loops can be nested as many times as necessary:

```
>>for i=0 :10, for j=1 :5, a=i*j; end; end
```

A1.5. Writing a *Matlab* script

Matlab programs are saved with the file extension ".m". When they are executed, error messages are displayed whenever *Matlab* encounters an error, indicating where in the code the error arose. Scripts do not need to be precompiled. To run a program, we must first load the file directory in which it is saved.

Data files are saved with the file extension ".mat", and variables are saved with double precision.

A1.6. Generating figures with *Matlab*

Matlab is a very powerful and user-friendly tool for managing figures, whether in one, two or three dimensions. For example, to plot the curve y=sin(x), where x=0:50, we can simply run:

```
>>x= -pi:pi/10:pi; y=sin(x), plot(x,y)
```

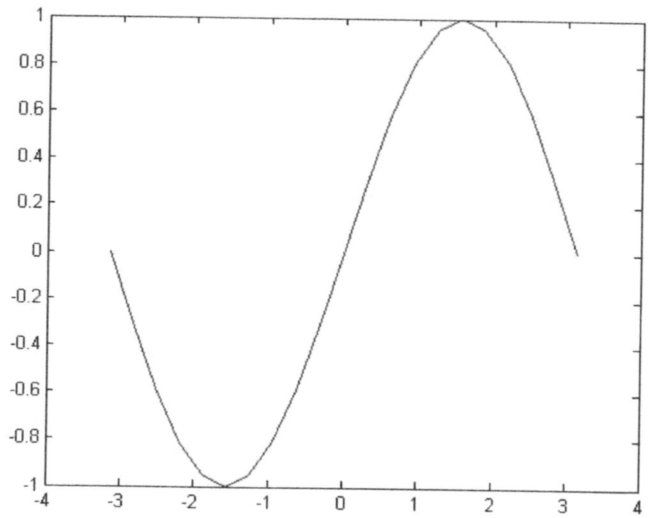

Some of the many commands for plotting graphs and manipulating axes and scales are briefly summarized below:

– xlabel("time"): gives a title to the x-axis;

– ylabel("speed"): gives a title to the y-axis;

– title("progression of the speed"): gives a title to the graph;

– text(2,4,"+++Temperature T1"): adds a caption to the curve plotted by "+++" at a certain point;

– loglog(x,y): plots the curve with a logarithmic scale (log-log);

– semilogx(t,f(t)): plots the curve with a logarithmic scale along the x-axis only;

– semilogy(t,f(t)): plots the curve with a logarithmic scale along the y-axis only;

– grid on: displays the grid in the graph;

– grid off: hides the grid in the graph;

– clf: deletes the graph;

– plot(x,y,x,z,x,w): plots y, z and w as a function of x on the same graph;

– polar(x,y): plots the curve y as a function of x in polar coordinates;

– plot(x,y,"+g"): plots y as a function of x using green "+" symbols;

– fplot("f_name",[x-min, x-max]): plots the function "f_name" on the specified interval of x;

– axis("square"): plots a square graph;

– axis("off"): hides the x- and y-axes;

– axis("on"): displays the x- and y-axes;

– axis([x-min, x-max, y-min, y-max]): displays the section of the graph between the specified values on the x- and y-axes;

– plot3(x,y,z): plots z in 3D as a function of x and y.

Appendix 2

General Approximation Principles

A2.1. Standard results

DEFINITION.– *A Hilbert space is a space equipped with a scalar product* $(.,.)_H$ *that is complete with respect to the norm induced by this scalar product.*

Let F be a closed vector subspace of a Hilbert space H. For every $x \in H$, there exists a unique $a \in F$ such that $||x - a||_H = d(x, F) = \inf_{b \in F} ||x - b||_H$.

This point $a = P_F x$ is said to be the projection of x onto F, which defines a function $P_F \colon H \to H$ satisfying $Im(P_F) = F$ and characterized by the property $(x - P_F x, b) = 0, \forall b \in F$.

The function P_F is known as the orthogonal projection of H onto F. For all $x, y \in H$:

$$||P_F x - P_F y||_H \leq ||x - y||_H.$$

DEFINITION.– *The bilinear form* $a(.,.)$ *and the linear form* l *are said to be continuous if and only if:*

$$\begin{array}{ll} \exists c_a > 0, \ \forall (u,v) \in V^2, & |a(u,v)| \leq c_a ||u||_V ||v||_V; \\ \exists c_l > 0, \ \forall v \in V, & |l(v)| \leq c_l ||v||_V. \end{array} \qquad [A2.1]$$

DEFINITION.– *The bilinear form* $a(.,.)$ *is said to be* α*-coercive if:*

$$\exists \alpha > 0, \quad \forall v \in V, \quad a(v,v) \geq \alpha ||v||_V^2. \qquad [A2.2]$$

The next theorem states that, in a Hilbert space, whenever a linear form l is continuous, it can be represented by a vector n that is normal to the kernel hyperplane $Ker(l) = l^{-1}(\{0\})$.

THEOREM.– *Let V be a Hilbert space with scalar product $(.,.)_V$ and let V' be its dual space (the space $V' = L(V, R)$ of linear forms that are continuous on V). Then:*

$$\forall l \in V', \quad \exists \tau_l \in V, \quad l(v) = (\tau_l, v)_V. \qquad [A2.3]$$

In other words, every continuous linear form on V can be viewed as an operator that takes the scalar product by some vector $\tau_l \in V$. Furthermore:

$$\|\tau_l\|_V = \|l\|_{V'}. \qquad [A2.4]$$

THEOREM.– *Suppose that:*

i) V is a Hilbert space with scalar product $(.,.)_V$ and norm $\|.\|_V$;

ii) $a(.,.)$ is a bilinear form that is α-coercive on V;

iii) $l(.)$ is a linear form that is continuous on V.

Then, the following problem is well-posed:

$$(P) \begin{cases} \textit{Find } u \in V \textit{ such that:} \\ a(u,v) = l(v) \end{cases}$$

In other words, the solution u exists, is unique and depends continuously on l:

$$\|u\|_V \leq \frac{\|l\|_{V'}}{\alpha}. \qquad [A2.5]$$

The bilinear form $a(.,.)$ can be identified with the operator A from V to its dual space $V' = L(V, \mathbb{R})$:

$$\forall u, v \in V, \quad < Au, v >_{V,V'} = a(u,v). \qquad [A2.6]$$

The problem (P) can therefore be reformulated as:

$$\begin{cases} \text{Find } u \in V \text{ such that:} \\ Au = l \ \text{ on } V' \end{cases} \qquad [A2.7]$$

This form [A2.7] of the problem (P) is known as the strong form. It gives an interpreted version of the problem.

In the examples considered below, the strong form corresponds to the PDEs that we wish to solve, whereas the variational formulation views the problem in terms of energy (*theorem of virtual power*). The Lax–Milgram theorem states that the operator $A^{-1} : V' \to V$ is continuous and bijective:

$$l \in V' \to u = A^{-1}l \in V, \quad [A2.8]$$

and

$$\|u\|_V = \|A^{-1}l\|_V \le c\|l\|_{V'}, \quad [A2.9]$$

where $c > 0$ is independent of $l : c = \frac{1}{\alpha}$. Therefore, a perturbation applied to l will produce an error of same order on u.

A2.2. Linear variational problems

In general, the variational formulation of a boundary value problem is of the following form:

$$(P)\begin{cases} \text{Find } u \text{in the Hilbert space } V \text{s.t.:} \\ \qquad a(u,v) = l(v) \quad v \in V \end{cases} \quad [A2.10]$$

with the assumptions:

– a is a continuous bilinear and elliptic form on V;

– l is a continuous linear form on V.

The Lax–Milgram theorem leads to the following conclusions:

– The problem P has a unique solution u on V.

– If the bilinear form a is symmetric, then the variational problem P is equivalent to the following minimization problem:

$$\begin{cases} \text{Find } u \in V \text{ that minimizes the quadratic form} \\ \qquad J(v) = \frac{1}{2}a(v,v) - l(v) \end{cases} \quad [A2.11]$$

A2.3. Variational approximation

Consider now a finite-dimensional Hilbert subspace $V_h \subset V$, indexed by h, equipped with the norm induced by V, and which satisfies the property that, for every $v \in V$, there exists an element $r_h v = v_h \in V_h$ such that:

$$\lim_{h \to 0} \|r_h v - v\|_V = 0.$$

This is said to be an internal approximation, because $V_h \subset V$, and V_h is a conformal approximation of V.

Thus, consider the problem P_h:

$$(P_h)\begin{cases} \text{Find the function } u_h \in V_h \text{ such that:} \\ \quad a(u_h, v_h) = l(v_h) \quad \forall v_h \in V_h \end{cases} \qquad [A2.12]$$

The problem (P_h) is well posed for any choice of h.

It also has a unique solution, because $V_h \subset V$, and the hypotheses of the Lax–Milgram theorem are also satisfied in V_h. Similarly, if the bilinear form a is symmetric, then the variational problem P_h is equivalent to the following minimization problem:

$$\begin{cases} \text{Find the function } u_h \in V_h \text{ that minimizes the quadratic form} \\ \qquad J(v_h) = \frac{1}{2}a(v_h, v_h) - l(v_h) \end{cases} \qquad [A2.13]$$

Clearly:

$$J(u_h) > J(u).$$

A2.4. General result on an upper bound for the error

THEOREM.– *Let M be the constant in the continuity hypothesis of a:*

$$a(u, v) \leq M\|u\|\|v\| \quad (M = \|a\|),$$

and let α be the constant in the ellipticity hypothesis:

$$a(v, v) \geq \alpha\|v\|^2.$$

Then, the following upper bound holds for the error:

$$\|u - u_h\| \leq \frac{M}{\alpha} \inf_{v_h \in V_h} \|u - v_h\|.$$

This is described as the a priori *approximation error.*

If a is also symmetric, then the fact that $a(u - u_h, v_h) = 0, \forall v_h \in V_h$ implies that u_h is the projection of u onto V_h with respect to the scalar product defined by a. In this case, we can give an improved *a priori* approximation:

$$\|u - u_h\| \leq \sqrt{\frac{M}{\alpha}} \inf_{v_h \in V_h} \|u - v_h\|.$$

By increasing the dimension of V_h, we might hope that u can be approximated "arbitrarily well". This will succeed whenever the V_h are contained in a dense subspace of W, i.e. it suffices to consider V_h spaces that are all of the same type, for example, all contained in the space W of continuous piecewise affine functions, which is known to be dense in $H^1(\Omega)$.

A2.5. Speed of convergence

Suppose that Ω is an open subset of $\mathbb{R}^n$ and that Ω is partitioned into N_{el} subdomains K_i (the elements or mesh cells):

$$\mathcal{T} = \cup_{i=1}^{N_{el}} K_i \text{ where } \forall i, j,\ K_i \cap K_j = \emptyset.$$

Let h_i be the diameter of the K_i (i.e. the diameter of the smallest ball containing K_i), and write $h = \max(h_i)$ for the diameter of the largest cell.

We will assume that the approximation is conformal and convergent, i.e. the approximation space V_h tends to V as h tends to 0. In other words, for arbitrary $u \in V$:

$$d(u, V_h) \to 0,$$

where $d(u, V_h) = \inf_{v_h \in V_h} \|u - v_h\|_V$.

It can be shown that approximating by continuous piecewise k-th order polynomials on Ω results in an approximation error of order h^k:

$$\|u - u_h\|_V \leq c_k h^k,$$

where $c_k > 0$ is a constant that is independent of h.

Hence, it might seem desirable to prefer higher-order polynomial approximations. However, the implementation (programming) will be more difficult and may increase the cost significantly for a marginal gain in accuracy.

A2.6. Galerkin method

Suppose that the Hilbert space V is separable. Then, there exists a Hilbert basis $\{w_j\}_{j=1}^{+\infty}$ that generates a dense subspace of V. Consider a finite subset $B_m = \{w_j\}_{j=1}^{m}$ and write V_m for the space generated by B_m.

If Π_m is the orthogonal projection vector from V to V_m, then:

$$\lim_{m \to +\infty} \|\Pi_m v - v\|_V = 0 \quad \forall v \in V_m.$$

Indeed, because every element can be written as $v = \sum_{i=1}^{+\infty} v_i w_i$, its projection onto V_m is $\Pi_m v = \sum_{i=1}^{m} v_i w_i$, and the result follows from the approximation theorem, given that $\|v - \Pi_m v\|_V \to 0$. Therefore, we need to find a solution of the following problem P_m:

$$(P_m) \begin{cases} \text{Find } u_m \in V_m \text{ such that:} \\ a(u_m, v_m) = l(v_m) \quad \forall v_m \in V_m \end{cases} \qquad \text{[A2.14]}$$

Whenever the Lax–Milgram theorem is satisfied, this problem is well posed and has a unique solution $u_m \in V_m$. It can then be shown that:

$$\|u - u_m\|_V \to 0.$$

We note that the approximate solution constructed in this way is “truncated”; we are neglecting the components of u with indices $\geq m$. This method is not used in practice, because we do not know a Hilbert basis of V in general. The methods that are used instead (non-Hilbert basis, subspaces V_m not contained in V_{m+l} for $l \in \mathbb{N}^*$) are nevertheless often referred to as Galerkin methods.

Bibliography

[ATL 98] ATLURI S.N., ZHU T., "A new meshless local Petrov-Galerkin approach in computational mechanics", *Computational Mechanics*, vol. 22, pp. 117–127, 1998.

[ATT 94] ATTAWAY S.W., HEINSTEIN M.W., SWEGLE J.W., "Coupling of smooth particle hydrodynamics with the finite element method", *Nuclear Engineering and Design*, vol. 150, pp. 199–205, 1994.

[BAK 76] BAKHVALOV N., *Méthodes numériques - Analyse, algèbre, équations différentielles ordinaires*, Mir Publishers, 2nd edition, 1976.

[BAT 96] BATHE K.-J., *Finite Element Procedures*, Prentice Hill, 1996.

[BEL 94] BELYTSCHKO T., LU Y.Y., GU L., "Element-free Galerkin methods", *International Journal for Numerical Methods in Engineering*, vol. 37, pp. 229–256, 1994.

[BEL 06] BELINHA J., DINIS L.M., "Elastoplastic analysis of plates by the element free Galerkin method", *International Journal of Computer Aided Engineering and Software*, vol. 23, no. 5, pp. 525–551, 2006.

[BEL 07] BELINHA J., DINIS L.M., "Nonlinear analysis of plates and laminates using the element free Galerkin method", *Composite Structures*, vol. 78, pp. 337–350, 2007.

[BIR 06] BIRD R.B., STEWART W.E., LIGHTFOOT E.N., *Transport Phenomena*, John Wiley & Sons, 2nd revised edition, Hoboken, New Jersey, 2006.

[BRE 83] BREZIS H., *Analyse fonctionnelle, Théorie et Applications*, Masson, Paris, 1983.

[CHE 96] CHEN J.S., PAN C., WU C.T. *et al.*, "Reproducing kernel particle methods for large deformation analysis of non-linear structures", *Computer Methods in Applied Mechanics and Engineering*, vol. 139, pp. 195–227, 1996.

[CHE 98] CHEN J.S., WANG H.P., "New boundary conditions treatments in meshfree computation of contact problems", *Computer Methods in Applied Mechanics and Engineering*, vol. 187, pp. 441–468, 1998.

[CHE 01] CHEN J.S., WU C.T., YOON Y., "A stabilized conforming nodal integration for Galerkin mesh-free methods", *International Journal for Numerical Methods in Engineering*, vol. 50, pp. 435–466, 2001.

[CHE 08] CHEN S.S., LIU Y.H., CEN Z.Z., "Lower bound shakedown analysis by using the element free Galerkin method and non-linear programming", *Computer Methods in Applied Mechanics and Engineering*, vol. 197, pp. 3911–3921, 2008.

[CHE 09] CHEN S.S., LIU Y.H., CEN Z.Z., "A time integration algorithm for linear transient analysis based on reproducing kernel method", *Computer Methods in Applied Mechanics and Engineering*, vol. 198, pp. 3361–3371, 2009.

[CIA 88] CIARLET P.-G., *Mathematical Elasticity I: Three-dimensional Elasticity*, North-Holland, Amsterdam, 1988.

[CIA 90] CIARLET P., LIONS J., *Handbook of Numerical Analysis: Finite difference methods (pt. 1). Solutions of equations in Rn (pt. 1)*, North-Holland, 1990.

[CIA 02] CIARLET P., *The Finite Element Method for Elliptic Problems*, Society for Industrial and Applied Mathematics, 2002.

[CRO 84] CROUZEIX M., MIGNOT A., *Analyse numérique des équations différentielles*, Masson, Paris, 1984.

[DHA 81] DHATT G., TOUZOT G., *Une présentation de la méthode des éléments finis*, Presses de l'Université Laval Québec, Maloine, Paris, 1981.

[DOL 99] DOLBOW J., BELYTSCHKO T., "Numerical integration of the Galerkin weak form in meshfree methods", *Computational Mechanics*, vol. 23, pp. 219–230, 1999.

[DUA 96] DUARTE C.A., ODEN J.T., "An hp adaptive method using clouds", *Computer Methods in Applied Mechanics and Engineering*, vol. 139, pp. 237–262, 1996.

[DUV 98] DUVAUT G., *Mécanique des milieux continus*, Dunod, Paris, 1998.

[FAS 07] FASSHAUER G.F., *Meshfree Approximation Methods with MATLAB*, World Scientific Publishing Co., Inc., New Jersey, 2007.

[FIS 07] FISH J., BELYTSCHKO T., *A First Course in Finite Elements*, John Wiley & Sons, Hoboken, New Jersey, 2007.

[GUN 97] GUNTHER F.C., LIU W.K., "Implementation of boundary conditions for meshless methods", *Computer Methods in Applied Mechanics and Engineering*, vol. 163, pp. 205–230, 1997.

[HAR 07] HARTMANN F., KATZ C., *Structural Analysis with Finite Elements*, Springer, New York, 2007.

[HEG 96] HEGEN D., "Element-free galerkin methods in combination with finite element approaches", *Computer Methods in Applied Mechanics and Engineering*, vol. 19, pp. 120–35, 1996.

[HUE 04] HUERTA A., BELYTSCHKO T., FERNANDEZ-MENDEZ S. *et al.*, "Meshfree Methods", *Encyclopedia of Computational Mechanics*, John Wiley & Sons, Hoboken, New Jersey, 2004.

[KAL 97] KALJEVIC I., SAIGAL S., "An improved element free Galerkin formulation", *International Journal for Numerical Methods in Engineering*, vol. 40, 1997.

[KAR 04] KARGARNOVIN M.H., TOUSSI H.E., FARIBORZ S.J., "Elasto-plastic element-free galerkin method", *Computational Mechanics*, vol. 33, pp. 206–214, 2004.

[KOK 09] KOKO J., *Calcul scientifique avec MATLAB*, Ellipses, Paris, 2009.

[LAN 81] LANCASTER P., SALKAUSKAS K., "Surfaces generated by moving least squares methods", *Mathematics of Computation*, vol. 37, pp. 141–158, 1981.

[LEV 02] LEVEQUE R.J., *Finite Volume Methods for Hyperbolic Problems*, Cambridge University Press, 1st edition, 2002.

[LIE 07] LIEW K.M., PENG L.X., "Geometric non-linear analysis of folded plate structures by the spline strip kernel particle method", *International Journal for Numerical Methods in Engineering*, vol. 71, pp. 1102–1133, 2007.

[LIU 95] LIU W.K., JUN S., ZHANG Y.F., "Reproducing kernel particle methods", *International Journal for Numerical Methods in Engineering*, vol. 20, pp. 1081–1106, 1995.

[LIU 96] LIU W.K., CHEN Y., JUN S. *et al.*, "Advances in multiple scale kernel particle methods", *Computational Mechanics*, vol. 18, no. 2, pp. 73–11, 1996.

[LIU 05] LIU T., LIU G., WANG Q., "An element-free galerkin-finite element coupling method for elasto-plastic contact problems", *Journal of Tribology*, vol. 128, pp. 1–9, 2005.

[LUC 77] LUCY L.B., "A numerical approach to the testing of the fission hypothesis", *The Astronomical Journal*, vol. 82, pp. 1013–1024, 1977.

[MAT 10] MATEJ A., Meshless Numerical Methods, Project, University of Nova Gorica, 2010.

[MAT 14] MATLAB®, "The Language of Technical Computing", The Mathworks®, New York, 2014.

[MEL 96] MELENK J.M., BABUSHKA I., "The partition of unity finite element method: Basic theory and applications", *Computer Methods in Applied Mechanics and Engineering*, vol. 139, pp. 289–314, 1996.

[MUK 97] MUKHERJEE Y.X., MUKHERJEE S., "The boundary node method for potential problems", *International Journal for Numerical Methods in Engineering*, vol. 40, pp. 797–815, 1997.

[NAY 92] NAYROLES B., TOUZOT G., VILLON P., "Generalizing the finite element method: Diffuse approximation and diffuse elements", *Computational Mechanics*, vol. 10, pp. 307–318, 1992.

[PAT 80] PATANKAR S., *Numerical Heat Transfer Fluid Flow*, McGraw-Hill, New York, 1980.

[PHU 07] PHU N.V., RABCZUK T., BORDAS S. *et al.*, "Meshless methods: a review and computer implementation aspects", *Mathematics and Computers in Simulation*, 2007.

[QUA 04] QUARTERONI A., SACCO R., SALERI F., *Méthodes numériques: Algorithmes, analyse et applications*, Springer, New York, 2004.

[QUA 08] QUARTERONI A.M., SALERI F., GERVASIO P., *Calcul Scientifique*, Springer Verlag, 2008.

[RAB 07] RABCZUK T., AREIAS P., BELYTSCHKO T., "A meshfree thin shell method for nonlinear dynamic fracture", *International Journal for Numerical Methods in Engineering*, vol. 72, no. 5, pp. 524–548, 2007.

[RAD 09] RADI B., EL HAMI A., *Eléments d'analyse*, Ellipses, Paris, 2009.

[RAD 10] RADI B., EL HAMI A., *Mathématiques numériques pour l'ingénieur*, Ellipses, Paris, Technosup edition, 2010.

[RAO 04] RAO B.N., RAHMAN S., "An enriched meshless method for non-linear fracture mechanics", *International Journal for Numerical Methods in Engineering*, vol. 59, pp. 197–223, 2004.

[RAP 05] RAPPAZ J., PICASSO M., *Introduction à l'analyse numérique*, Presses polytechniques et universitaires romandes, Lausanne, 2005.

[SMA 02] SMAOUI H., RADI B., "Comparative study of different advective schemes: Application to the MECCA model", *Environmental Fluid Mechanics*, vol. 1, no. 4, pp. 361–381, 2002.

[TRE 96] TREFETHEN L.N., *Finite Difference and Spectral Methods for Ordinary and Partial Differential Equations*, Cornell University, New York, 1996.

[WAG 00] WAGNER G.J., LIU W.K., "Application of essential boundary conditions in meshfree methods: a corrected collocation method", *International Journal for Numerical Methods in Engineering*, vol. 47, pp. 1367–1379, 2000.

[XU 98] XU Y., SAIGAL S., "Element free galerkin study of steady quasi-static crack growth in plane strain tension in elastic-plastic materials", *Computational Mechanics*, vol. 22, pp. 255–265, 1998.

[XU 99] XU Y., SAIGAL S., "An element-free galerkin analysis of steady dynamic growth of a mode i crack in elastic-plastic materials", *International Journal of Solids and Structures*, vol. 36, pp. 1045–1079, 1999.

[ZHU 98] ZHU T., ATLURI S., "A modified collocation method and a penalty formulation for enforcing the essential boundary conditions in the element free Galerkin method", *International Journal for Numerical Methods in Engineering*, vol. 21, 1998.

[ZIE 00] ZIENKIEWICZ O., TAYLOR R., *The Finite Element Method: The Basis*, vol. I, Butterworth-Heinemann, Oxford, 5th edition, 2000.

Index

S, V, W

Other titles from

iSTE

in

Mechanical Engineering and Solid Mechanics

2018

BOREL Michel, VÉNIZÉLOS Georges
Movement Equations 4: Equilibriums and Small Movements
(Non-deformable Solid Mechanics Set – Volume 4)

RADI Bouchaib, EL HAMI Abdelkhalak
Advanced Numerical Methods with Matlab® 1: Function Approximation and System Resolution
(Mathematical and Mechanical Engineering SET – Volume 6)

SALENÇON Jean
Virtual Work Approach to Mechanical Modeling

2017

BOREL Michel, VÉNIZÉLOS Georges
Movement Equations 2: Mathematical and Methodological Supplements
(Non-deformable Solid Mechanics Set – Volume 2)
Movement Equations 3: Dynamics and Fundamental Principle
(Non-deformable Solid Mechanics Set – Volume 3)

BOUVET Christophe
Mechanics of Aeronautical Solids, Materials and Structures
Mechanics of Aeronautical Composite Materials

BRANCHERIE Delphine, FEISSEL Pierre, BOUVIER Salima,
IBRAHIMBEGOVIC Adnan
From Microstructure Investigations to Multiscale Modeling: Bridging the Gap

CHEBEL-MORELLO Brigitte, NICOD Jean-Marc, VARNIER Christophe
From Prognostics and Health Systems Management to Predictive Maintenance 2: Knowledge, Traceability and Decision
(Reliability of Multiphysical Systems Set – Volume 7)

EL HAMI Abdelkhalak, RADI Bouchaib
Dynamics of Large Structures and Inverse Problems
(Mathematical and Mechanical Engineering Set – Volume 5)
Fluid-Structure Interactions and Uncertainties: Ansys and Fluent Tools
(Reliability of Multiphysical Systems Set – Volume 6)

KHARMANDA Ghias, EL HAMI Abdelkhalak
Biomechanics: Optimization, Uncertainties and Reliability
(Reliability of Multiphysical Systems Set – Volume 5)

LEDOUX Michel, EL HAMI Abdelkhalak
Compressible Flow Propulsion and Digital Approaches in Fluid Mechanics
(Mathematical and Mechanical Engineering Set – Volume 4)
Fluid Mechanics: Analytical Methods
(Mathematical and Mechanical Engineering Set – Volume 3)

MORI Yvon
Mechanical Vibrations: Applications to Equipment

2016

BOREL Michel, VÉNIZÉLOS Georges
Movement Equations 1: Location, Kinematics and Kinetics
(Non-deformable Solid Mechanics Set – Volume 1)

BOYARD Nicolas
Heat Transfer in Polymer Composite Materials

CARDON Alain, ITMI Mhamed
New Autonomous Systems
(Reliability of Multiphysical Systems Set – Volume 1)

DAHOO Pierre Richard, POUGNET Philippe, EL HAMI Abdelkhalak
Nanometer-scale Defect Detection Using Polarized Light
(Reliability of Multiphysical Systems Set – Volume 2)

DE SAXCÉ Géry, VALLÉE Claude
Galilean Mechanics and Thermodynamics of Continua

DORMIEUX Luc, KONDO Djimédo
Micromechanics of Fracture and Damage
(Micromechanics Set – Volume 1)

EL HAMI Abdelkhalak, RADI Bouchaib
Stochastic Dynamics of Structures
(Mathematical and Mechanical Engineering Set – Volume 2)

GOURIVEAU Rafael, MEDJAHER Kamal, ZERHOUNI Noureddine
From Prognostics and Health Systems Management to Predictive Maintenance 1: Monitoring and Prognostics
(Reliability of Multiphysical Systems Set – Volume 4)

KHARMANDA Ghias, EL HAMI Abdelkhalak
Reliability in Biomechanics
(Reliability of Multiphysical Systems Set –Volume 3)

MOLIMARD Jérôme
Experimental Mechanics of Solids and Structures

RADI Bouchaib, EL HAMI Abdelkhalak
Material Forming Processes: Simulation, Drawing, Hydroforming and Additive Manufacturing
(Mathematical and Mechanical Engineering Set – Volume 1)

2015

KARLIČIĆ Danilo, MURMU Tony, ADHIKARI Sondipon, MCCARTHY Michael
Non-local Structural Mechanics

SAB Karam, LEBÉE Arthur
Homogenization of Heterogeneous Thin and Thick Plates

2014

ATANACKOVIC M. Teodor, PILIPOVIC Stevan, STANKOVIC Bogoljub, ZORICA Dusan
Fractional Calculus with Applications in Mechanics: Vibrations and Diffusion Processes

ATANACKOVIC M. Teodor, PILIPOVIC Stevan, STANKOVIC Bogoljub, ZORICA Dusan
Fractional Calculus with Applications in Mechanics: Wave Propagation, Impact and Variational Principles

CIBLAC Thierry, MOREL Jean-Claude
Sustainable Masonry: Stability and Behavior of Structures

ILANKO Sinniah, MONTERRUBIO Luis E., MOCHIDA Yusuke
The Rayleigh–Ritz Method for Structural Analysis

LALANNE Christian
Mechanical Vibration and Shock Analysis – 5-volume series – 3rd edition
Sinusoidal Vibration – Volume 1
Mechanical Shock – Volume 2
Random Vibration – Volume 3
Fatigue Damage – Volume 4
Specification Development – Volume 5

LEMAIRE Maurice
Uncertainty and Mechanics

2013

ADHIKARI Sondipon
Structural Dynamic Analysis with Generalized Damping Models: Analysis

ADHIKARI Sondipon
Structural Dynamic Analysis with Generalized Damping Models: Identification

BAILLY Patrice
Materials and Structures under Shock and Impact

BASTIEN Jérôme, BERNARDIN Frédéric, LAMARQUE Claude-Henri
Non-smooth Deterministic or Stochastic Discrete Dynamical Systems: Applications to Models with Friction or Impact

EL HAMI Abdelkhalak, RADI Bouchaib
Uncertainty and Optimization in Structural Mechanics

KIRILLOV Oleg N., PELINOVSKY Dmitry E.
Nonlinear Physical Systems: Spectral Analysis, Stability and Bifurcations

LUONGO Angelo, ZULLI Daniele
Mathematical Models of Beams and Cables

SALENÇON Jean
Yield Design

2012

DAVIM J. Paulo
Mechanical Engineering Education

DUPEUX Michel, BRACCINI Muriel
Mechanics of Solid Interfaces

ELISHAKOFF Isaac *et al.*
Carbon Nanotubes and Nanosensors: Vibration, Buckling and Ballistic Impact

GRÉDIAC Michel, HILD François
Full-Field Measurements and Identification in Solid Mechanics

GROUS Ammar
Fracture Mechanics – 3-volume series
Analysis of Reliability and Quality Control – Volume 1
Applied Reliability – Volume 2
Applied Quality Control – Volume 3

RECHO Naman
Fracture Mechanics and Crack Growth

2011

KRYSINSKI Tomasz, MALBURET François
Mechanical Instability

SOUSTELLE Michel
An Introduction to Chemical Kinetics

2010

BREITKOPF Piotr, FILOMENO COELHO Rajan
Multidisciplinary Design Optimization in Computational Mechanics

DAVIM J. Paulo
Biotribolgy

PAULTRE Patrick
Dynamics of Structures

SOUSTELLE Michel
Handbook of Heterogenous Kinetics

2009

BERLIOZ Alain, TROMPETTE Philippe
Solid Mechanics using the Finite Element Method

LEMAIRE Maurice
Structural Reliability

2007

GIRARD Alain, ROY Nicolas
Structural Dynamics in Industry

GUINEBRETIÈRE René
X-ray Diffraction by Polycrystalline Materials

KRYSINSKI Tomasz, MALBURET François
Mechanical Vibrations

KUNDU Tribikram
Advanced Ultrasonic Methods for Material and Structure Inspection

SIH George C. *et al.*
Particle and Continuum Aspects of Mesomechanics

Printed and bound by CPI Group (UK) Ltd, Croydon, CR0 4YY
16/09/2026
14963181-0001